D1448147

PSYCHIATRY
AND GENETICS

PSYCHIATRY AND GENETICS

Psychosocial, Ethical, and Legal Considerations

EDITED BY

Michael A. Sperber, M.D.

Lissy F. Jarvik, M.D., Ph.D.

Basic Books, Inc., Publishers NEW YORK

ACKNOWLEDGMENTS

The editors acknowledge, with grateful appreciation, the participation of a number of individuals and institutions who made this book possible: Shervert H. Frazier, Jr., M.D., Professor of Psychiatry, Harvard Medical School, and Psychiatrist-in-Chief, McLean Hospital, suggested the editorial collaboration and proposed a number of the contributors. The National Foundation-March of Dimes provided a grant for the initial symposium at which many of the papers in this volume were originally presented, and its Vice-President for Professional Education, Daniel Bergsma, M.D., M.P.H. steadfastly supported the book's plan from its inception. Herb Reich, Director, Behavioral Sciences Program, Basic Books, has been a model of patient tact and gracious good will throughout our endeavors. Delia N. Daniels, N.H. Hancock, and Natalie Paul, Assistant Editor, The National Foundation, provided valuable editorial assistance. To them, and to our families and their progeny, this book is dedicated.

Library of Congress Cataloging in Publication Data
Main entry under title:

Psychiatry and genetics.

Includes bibliographical references and index.
1. Mental illness—Genetic aspects. 2. Genetic counseling. I. [DNLM: 1. Genetic intervention. 2. Genetic counseling. 3. Ethics, Medical. 4. Psychopathology. WM100 P986]
RC455.4.G4P78 616.8'9'042 73-36770
ISBN 0-465-06461-2

Printed in the United States of America
76 77 78 79 80 10 9 8 7 6 5 4 3 2 1

CONTENTS

PART II
Genetic Counseling

CONTRIBUTORS

Julie Anne Brody, M.A.
Department of Clinical Psychology
George Washington University
Washington, D.C.

Remi J. Cadoret, M.D.
Professor of Psychiatry
University of Iowa Medical School
Iowa City, Iowa

Barbara F. Crandall, M.D.
Assistant Professor of Pediatrics and Psychiatry
UCLA
Los Angeles, California

Park S. Gerald, M.D.
Professor of Pediatric Medicine
Harvard Medical School, and Chief, Division of Clinical Genetics
Children's Hospital Medical Center
Boston, Massachusetts

Lissy F. Jarvik, M.D., Ph.D
Professor of Psychiatry
UCLA, and Chief, Psychogenetics Unit
Veteran's Administration Hospital, Brentwood
Los Angeles, California

Marc Lappé, Ph.D.
Associate for the Biological Sciences
Institute of Society, Ethics and the Life Sciences
Hastings-on-Hudson, New York

Edward H. Liston, M.D.
Assistant Professor of Psychiatry
UCLA
Los Angeles, California

Albert S. Moraczewski, O.P., Ph.D.
President, Pope John XXIII Medical-Moral Research and Education Center
St. Louis, Missouri

John D. Rainer, M.D.
Chief, Psychiatric Research (Medical Genetics)
New York State Psychiatric Institute, and Professor of Clinical Psychiatry, College of Physicians and Surgeons
Columbia University
New York, New York

Julius B. Richmond, M.D.
Professor of Child Health and Human Development
Harvard Medical School
Director, Judge Baker Guidance Center
Boston, Massachusetts

Margery W. Shaw, M.D.
Professor of Medical Genetics
University of Texas
School of Biomedical Sciences
Houston, Texas

Michael A. Sperber, M.D.
Clinical Instructor in Psychiatry
Harvard Medical School, and Assistant Attending Physician
McLean Hospital
Belmont, Massachusetts

Robert J. Stoller, M.D.
Professor of Psychiatry
UCLA
Los Angeles, California

George Tarjan, M.D.
Professor of Psychiatry
UCLA, and Director, Mental Retardation and Child Psychiatry Program
Neuropsychiatric Institute, UCLA
Los Angeles, California

Stanley Walzer, M.D.
Assistant Professor of Psychiatry
Harvard Medical School, and Senior Associate in Clinical Genetics
Children's Hospital Medical Center
Boston, Massachusetts

George Winokur, M.D.
P. W. Penningroth Professor and Chairman, Department of Psychiatry
University of Iowa Medical School, and Psychiatrist-in-Chief
Psychopathic Hospital
Iowa City, Iowa

PREFACE

Innovations in genetic technology and the changing moral and legal climate surrounding amniocentesis and abortion now make it possible for mankind to participate more than ever before in shaping the genetic composition of future generations. The ethical, legal, and social implications of this development have caused grave concern to genetic counselors* but have remained largely outside the purview of psychiatrists. Indeed, with a few notable exceptions, psychiatrists have generally tended to ignore the field of human genetics as such. There are compelling reasons why they can no longer afford to do so. First, genetic influences upon behavior are being uncovered. Second, psychiatrists are expected to advise patients and their relatives as to the role of genetic factors in psychopathology. In addition, psychiatrists, like psychologists, can evaluate more adequately than most others in the medical profession the intrapsychic effects of such events as abortion, artificial insemination, the birth of an abnormal child, or the revelation that one is the carrier of an undesirable trait. Psychiatrists have no special expertise as far as the ethical, legal, social, and moral issues themselves are concerned. They do, however, by virtue of their professional training, have special skills in understanding the personal meanings of these issues and can therefore contribute a dimension much needed by the individual.

* See B. Hilton, M. Harris, P. Condliffe, and V. Berkley (eds.): *Ethical Issues in Human Genetics: Genetic Counseling and the Use of Genetic Knowledge* (New York: Plenum Press, 1973).

Finally, if the effectiveness in interpersonal communication displayed by psychiatrists in the therapeutic setting can be transferred to the counseling situation, participation of psychiatrists in genetic counseling may help reduce significantly the misunderstanding and lack of understanding that are so frequently the result of genetic counseling.

In order to promote dialogue between psychiatrists, geneticists, and ethicists, The National Foundation sponsored a panel entitled *Genetic Counseling: Psychiatric, Legal, Ethical, and Theologic Issues* at the 126th annual meeting of the American Psychiatric Association, May 10, 1973. The present volume is an outgrowth of that panel discussion, several of the papers having been expanded and others added for publication.

The papers in the first section, *Psychopathology and Genetics,* are intended to provide an overview of modes of transmission and to sketch present knowledge concerning the various categories of psychopathology. In the second section, *Genetic Counseling,* intrapsychic, ethical, moral, and legal issues are discussed. The division has been somewhat arbitrary, since several of the authors in the first section address themselves to these issues too. The very first chapter, for example, points to the complex ethical, moral, and social questions raised by conditions such as Huntington chorea and extra sex chromosomes. And Stoller concludes his chapter on gender disorders by warning that, with rare exceptions, "one's knowledge of the genetic state should play no part in treatment decisions." In the second section, Lappé and Brody stress the potential danger of genetic counseling as a form of "behavior control" in its pejorative sense, while Walzer and colleagues discuss the need for long-term contact with genetic counselors to avoid the misunderstandings, distortions, and denials which so often accompany the communication of affect-laden material. In the final chapters, we have Moraczewski's advocacy of the right to life—the case against abortion—and Shaw's opinion that the time may come when parents will be sued for "wrongful life" by children willfully and knowingly conceived despite information available that "the risk of genetic disease or defect is high and that the burden of the disease is great."

Depending on their own cultural, religious, social, and familial origins, as well as their personal lifelong experiences, readers will tend to adopt one or another of the views espoused in the succeeding chapters. After all, readers, whether or not they are psychiatrists, are not immune to the mechanisms determining behavior in all of us, and their opinions encompass the extremes of the continuum ranging from social control to social activism. Those psychiatrists who strongly favor social control tend to view mental illness as organismic and are inclined to effect autoplastic changes that will enable the patient to adapt to, and cope with, society as he finds it. By contrast, the social activist first considers the possibility of eliminating existing social pathogens, since he believes that factors external to the individual are the crucial issues in the etiology of mental disorders. Perusal of the chapters in this volume will make it clear that there is no need for an either/or position, that there is continued interaction between genotype and psychosocial forces, that in certain instances genetic predisposition may be equated with therapeutic optimism and even prophylactic success (e.g., lithium prophylaxis in bipolar disease), and that, as with all technologies, genetic and psychotechnology is neither good nor bad, but derives its value from those who use it.

Clearly, then, it is imperative that genetic counselors identify and make explicit underlying assumptions and ethical constructions, and that they develop an appreciation of the ethical, social, and legal complexities at the interface between psychiatry and genetics. Ignoring these complexities could have tragic consequences—the atrocious example of eugenics in Nazi Germany is still vividly within the minds of many of us and should serve as a warning for a long time to come.

The underlying issues are complex and thorny. Basically they involve a tension between the consumer's rights (to privacy; to know or not to know; to treatment or to the refusal of treatment) and society's needs (to know; to provide quality care to maximum numbers). This tension must be reduced in a way which avoids abuse or exploitation of both the individual and society.

There are no facile solutions. How, for example, can one meaningfully inform an individual of the risks involved in mass genetic screening when these risks, if they exist, have not yet been determined?

One of medicine's old maxims comes to mind as innovations in psychogenetic technology multiply—*primum non nocere!*

Michael A. Sperber, M.D.
Lissy F. Jarvik, M.D., Ph.D.

EDITORIAL COMMENT

The reader will note that the possessive has been deleted from eponyms throughout this volume; for example, Down syndrome, Huntington chorea, Hurler syndrome. This new format is in accord with the guidelines set at a meeting held February 10–11, 1975, at the National Institutes of Health and attended by internationally outstanding individuals in the field. In their report, entitled "Proposed Guidelines for the Classification, Nomenclature, and Naming of Morphologic Defects," item No. IV–D of the proposals for naming patterns of malformations states, "The possessive use of an eponym should be discontinued, since the author neither had nor owned the disorder" (*J. Pediatr.* 87: 162–164, 1975).

PART I

Psychopathology and Genetics

1

GENETIC MODES OF TRANSMISSION RELEVANT TO PSYCHOPATHOLOGY *

Lissy F. Jarvik, M.D., Ph.D.

Introduction

That in our era of technocracy a book devoted to genetics—the scientific discipline so intimately identified with technologic breakthroughs—should be deeply concerned with questions of ethics, morals, and laws is a testimony to the flexibility of the human spirit. Equally, our preoccupation with the selfsame questions that have tortured philosophers and theologians since the dawn of recorded history is a powerful argument in support of the rigidity of those mental processes that are distinguishing attributes of Homo sapiens.

And yet, there have always been those who would ignore the biologic determinants of behavior, as exemplified by Watson's famous dictum: "Give me a dozen healthy infants, well-formed, and my own specified world to bring them up in and I'll guarantee to take any one at random and train him to become . . . doctor, lawyer, artist, merchant-chief, and yes, even beggarman and thief . . ." [1] (p.82). Surely, with all of

* The generous assistance of Dr. Steven S. Matsuyama in the preparation of the manuscript, the illustrations, and the reference list is hereby gratefully acknowledged.

the evidence accumulated in the past 50 years concerning the influence of genetic factors on human behavior, we should be able to assume that the Watsonian point of view has become obsolete. It has not.

Lack of adequate dissemination of the information gathered by behavior geneticists, together with the erroneous assumption that a genetic predisposition is unalterable because it is inherited, is probably responsible for the persisting prevalence of the belief that heritability equals incurability and unmodifiability. This belief prevails despite the fact that, aside from infectious diseases and nutritional deficiencies, prevention and cure have been possible primarily in disorders identified as being of genetic origin. In the field of mental retardation, for example, there is little we can do for victims of birth trauma, but we can prevent the devastating consequences of genetically determined disorders in thyroid, galactose, and phenylalanine metabolism, to mention but a few. Indeed, it is not unlikely that once sufficient understanding of biobehavioral interactions has been gained, a condition will be considered curable, modifiable, and preventable whenever a significant heritable component can be defined.

It is the purpose of this chapter to present some of the accumulated data on behavior genetics. The review is by no means all-inclusive, or even comprehensive, the area of mental retardation constituting by far the largest omission (discussed in the chapter by Crandall and Tarjan). Rather, it is designed to highlight certain basic genetic principles as they apply to psychopathology.

Gene-borne Abnormalities

AUTOSOMAL DOMINANT: HUNTINGTON CHOREA

The oldest and best known genetic concept is that of single-gene inheritance, discovered by the Bohemian monk Gregor Mendel in his garden peas over 100 years ago [2] but ignored until it was rediscovered independently in 1900 by an Austrian, E. V. Tschermak,[3] a Dutchman, H. de Vries,[4] and a German, C. Correns.[5] A classic example of single-gene inheritance in man is Huntington chorea, identified as a hereditary disorder almost two decades before Mendel first reported his experiments. Although Waters's report [6] antedated that of Huntington, it was the latter's accurate description of the disease and its mode of transmission that became the classic reference.[7] Huntington chorea is characterized by the onset, usually in adulthood, of progressively worsening choreiform movements, dementia, and loss of cells in the caudate nucleus, putamen, and cerebral cortex, particularly the frontal lobes.

The hereditary pattern of Huntington chorea follows an autosomal dominant mode of transmission. A change that has taken place in a small region of a single DNA molecule (a point mutation) in one of the 22 pairs of autosomes (chromosomes that are not sex chromosomes) is transmitted from parent to offspring. This mutation, occurring in only one of the paired genes (or alleles), is sufficient to cause the disease, despite the presence of the other, normal gene. Hence, the mutant gene is said to be dominant over the normal gene. The typical features of dominant inheritance were reported by Huntington, who pointed out that if either parent manifested the disease, one or more of the offspring invariably succumbed—provided they lived long enough. If the offspring go through life without manifesting its effects, however, "the thread is broken and the grandchildren or great-grandchildren of the

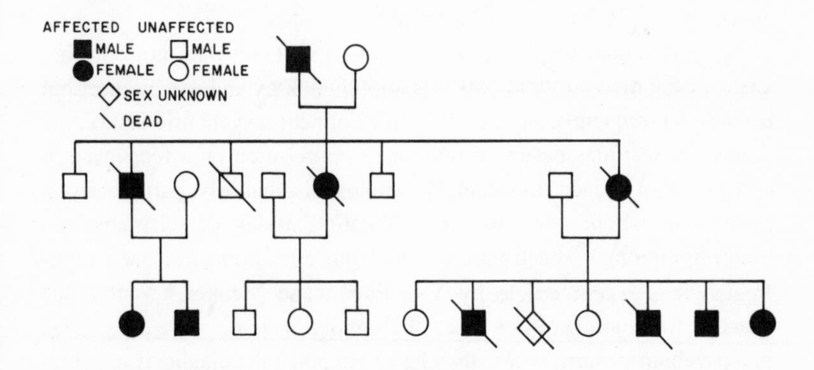

Figure 1. Pedigree of a family with Huntington chorea.
Source: K. W. G. Heathfield and I. C. K. Mackenzie, Huntington's chorea in Bedfordshire, England. *Guys Hosp. Rep. 120:*295, 1971. Reprinted by permission.

original Shakers may rest assured that they are free from the disease. . . . Unstable and whimsical as the disease may be in other respects, in this it is firm, it never skips a generation to manifest itself in another." [8]

A pedigree illustrating the key features of the transmission of Huntington chorea is shown in Figure 1. (1) Only offspring of affected individuals can inherit the trait. Descendants of nonaffected individuals are safe. (2) On the average, 50 percent of the offspring of affected individuals develop the disease. In this particular pedigree, 9 out of 18 descendants are affected. (3) It is impossible to predict for any given individual how many of his or her offspring will develop the disease, since each offspring has a 50–50 chance, regardless of how many other offspring are either affected or free of the disease. Thus, we see in the third generation that whereas in one sibship of 2, both individuals are affected; there is another sibship of 3 where none are affected; and a third sibship where 4 out of 7 are affected. (4) Both sexes are affected equally—the ratio of male to female (5 to 4) approximates 1:1.

Huntington chorea, though classic in every other respect, has one feature that imposes a special burden upon both the genetic counselor and the potential carrier of the disease. The disease onset is so late that it is frequently first noticeable at a time when potential carriers have already had their families. Although the age of onset is most commonly between 35 and 45 years, the range is enormous, and cases as young as 5 and as old as 70 have been recorded.

The differential diagnosis may be a very difficult one, not only because of the marked variations in symptomatology and age at onset but because so frequently any family history of neuropsychiatric disorder is denied. Sometimes based on ignorance, more often on a fear-inspired conspiracy of silence, this denial promotes misdiagnosis, particularly in patients in whom psychotropic medication masks the characteristic motor symptoms. The frequency of diagnostic errors has been highlighted in a recent article by Van Putten and Menkes.[9] Within the space of four months, three cases of Huntington chorea were discovered in a psychiatric ward, where they had been given the diagnosis of schizophrenia and later "schizophrenia with a phenothiazine-induced movement disorder." Incidentally, all of these patients had neurologic consultations which resulted in the additional diagnosis of tardive dyskinesia. As the authors point out, "although there is no satisfactory treatment for Huntington's Disease, prompt diagnosis is important to allow appropriate genetic counseling. Furthermore, the presumed drug induced dyskinesia is often treated with anticholinergic preparations which make Huntington's chorea worse." [10]

As of this writing, there is no way to distinguish, prior to the appearance of clinical symptoms, those persons who will develop Huntington chorea from those who will not. This state of affairs is generally considered unfortunate, for we cannot grant relief from the terror of developing this illness to those who are to be spared and who are fully aware, by virtue of their family history, of what may be in store for them. "It is spoken of by those in whose brain the seeds of the disease are known to exist, with a kind of horror and not at all alluded to except through dire necessity, when it is mentioned as 'that disorder' " (p. 298 [8]). Not only are we unable to detect carriers of the gene for Huntington chorea before they manifest obvious symptomatology, but we are also unable to prevent, arrest, or change the course of the disease.

In view of our total impotence, then, is it not fortunate, rather than unfortunate, that we are unable to detect the carriers? Our inability to do so may not continue too long if ongoing investigations confirm preliminary reports of differential results for carriers on measurements of motor control,[11] response to L-dopa loading,[12] and ceruloplasmin level. Even approaches that were unsuccessful in the past, such as electroencephalographic analyses and assessment of premorbid personality and cognitive development, are being reexplored in the belief that advances in techniques and research design may help to uncover hitherto hidden deviations.

Once we are able to detect asymptomatic carriers, should we tell them? All of them? Some of them? Is the reassurance we would be able to give to those who are not carriers of the gene, that neither they nor their descendants will be afflicted, worth the certainty of doom that the carriers would simultaneously experience? Huntington chorea, unlike most other diseases, appears to be fully penetrant—every carrier of the gene will manifest the disease, unless he dies of some other cause first. Is it not better to have all the offspring of a Huntington family live, even if under the suspended sword of Damocles, in the hope that they might escape, rather than to free certain ones of the fear and confirm for others the inescapability of this dreaded fate? After all, given hope, man's adaptability to adversity is astounding. And yet, can we don the mantle of divinity and deny knowledge of the ultimate end to those whose predestined mode of dying we have discovered? Can we hide from another human being information that may be vital in his choice of the way he lives?

Are not these the very questions daily faced by physicians when caring for patients afflicted with an incurable illness? Is the problem of the psychiatrist any different from that of the internist deliberating how and to whom to communicate a newly made diagnosis of bronchogenic carcinoma? Yes, there is a vital difference: personal and social tragedies of transmission to future generations apply to Huntington chorea in a clear-cut, unequivocal manner, but generally not to neoplasia. While certain hereditary factors have been implicated in malignant tumors, the mode of transmission remains unknown.

One might expect that families with Huntington chorea would deliberately curtail procreation. The sibs of patients with Huntington chorea do fulfill this expectation. Compared to the population at large, they have a lower reproductive rate, 80 percent according to one informed estimate. By contrast, there are a number of studies showing that the patients themselves have a reproductive rate that ranges from slightly over one to nearly two times that of their unaffected sibs.

A single dominant mode of transmission is the example par excellence for demonstrating the effectiveness of reproductive control. If all descendants of Huntington chorea families were to refrain from bearing offspring, the frequency of the disorder could be reduced in one generation to that produced by spontaneous mutations. The net result would be an 8- to 14-fold reduction from the present rate of 4–and 7/100,000 down to 0.5/100,000.

Although reproductive control has a potent impact upon gene frequency in an autosomal dominant trait like Huntington chorea, reduction in the fertility of individuals affected by recessive disorders would make an insignificant contribution to the reduction of the gene pool, since the overwhelming reservoir of harmful genes resides in phenotypically normal heterozygotes. The prime example is that of phenylketonuria, and there are many others in the field of mental retardation.

AUTOSOMAL RECESSIVE: WILSON DISEASE

Even in the field of psychopathology—as distinguished from mental retardation—there is an example of single recessive inheritance. Progressive hepatolenticular degeneration, first described by Wilson in 1912,[13] is characterized by extrapyramidal symptoms, often found in association with mental changes. It results from a single gene defect transmitted by the recessive mode of inheritance (Figure 2). The marriage of a person suffering from Wilson disease to a normal partner, in whose ancestry there was no intermarriage with carriers of the disease, produces apparently normal offspring. Hence, all the children resemble in appearance (phenotype) the parent carrying the normal, dominant trait. None of the children will develop the disease—yet all of them will be carriers of the recessive, phenotypically hidden trait. In their genetic make-up (genotype) the children resemble neither parent. It is only among the grandchildren that the parental genotypes recur. If an offspring marries a genotypically normal spouse, one who does not carry the gene for Wilson disease, then none of their offspring will become affected, although half of them will become carriers. The other half will be normal genotypically as well as phenotypically, like their normal grandparent. If the offspring of the original marriage marries a heterozygous spouse like himself, the familiar Mendelian ratios appear. One quarter of the offspring will be completely normal, genotypically as well as phenotypically like the normal grandparent; one quarter will be affected, being homozygous for the recessive gene which is present in double dose; and one half will be heterozygous carriers, phenotypically normal like their parents. The third possibility is that a carrier marries a homozygous recessive individual—and in Wilson disease that is possible, since the onset of the disease is sometimes later in life (the manifestation period ranges from 6 to 40 years, with a peak at age 13). In that case, there will be no genotypically normal offspring, half of them being carriers and phenotypically normal, the other half being homozygous recessive and phenotypically affected. In Wilson disease the dictum of "skipping a generation" is well illustrated, since the disease skips from grandparent to grandchild.

In Wilson disease, unlike Huntington chorea, we know that the effect of the aberrant gene (when homozygous) is a deficiency of a specific

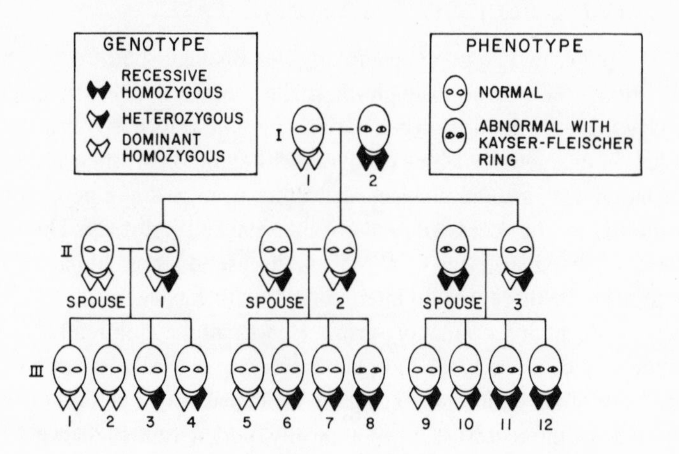

Figure 2. Theoretic pedigree of Wilson disease.
Source: L. F. Jarvik, Genetic aspects of aging. In Rossman, I. (ed.): *Clinical Geriatrics*. Philadelphia: J. B. Lippincott, 1971, p. 85. Reprinted by permission.

plasma protein (ceruloplasmin). This deficiency, in turn, is responsible for the accumulation of excess copper, with a consequent deposition of the copper in the liver parenchyma, the basal ganglia, and the cornea—the latter producing the pathognomonic Kayser-Fleischer ring. Moreover, as genetic counselors we do not face the same dilemma with Wilson disease as with Huntington chorea, i.e., the possibility of carrier detection without the accompanying ability to alter the course of the disease. In Wilson disease, homozygotes can be detected prior to the onset of symptoms, and penicillamine, together with dietary management, have been therapeutically successful. If introduced in childhood, it may prove efficacious in prophylaxis. Clearly, it is essential to identify carriers and, as soon as reliable techniques become available, to divulge the pertinent information to all individuals, affected and nonaffected. Those who are not carriers can be relieved of all anxiety; those who are homozygous for the trait can be treated prophylactically; and those who are heterozygous carriers can be assured that their children can receive prophylaxis and, therefore, will not be doomed to suffer the deleterious consequences of their inheritance. We face a different dilemma, however, as far as society as a whole is concerned. Patients with arrested or controlled disease can be expected to reproduce, and even though the overall frequency of the disease is low now (gene frequency 1/200–500), with longer survival and increased chances for reproduction, Wilson disease is likely to become more common in future generations. Should one advocate sterility for heterozygous as well as homozygous carriers of the genes? Were such a program instituted, it would be possible to reduce the frequency of the disease to that of the spontaneous mutation rate within a single generation.

X-LINKED RECESSIVE: LESCH-NYHAN SYNDROME

Both Huntington chorea and Wilson disease affect the sexes equally. There is another form of hereditary transmission commonly observed in man: X-linked transmission. X-linked recessive disorders are a special category. Although the name recessive implies that expression of the abnormality requires the presence of two mutant alleles, location of the mutant gene on the X chromosome restricts homozygosity to females only, since females have two X chromosomes. Males, having only a single X chromosome, require but a single dose of the mutant gene to manifest the disorder. Moreover, since all males receive their Y chromosome from the father and their X chromosome from the mother, X-linked traits are of necessity transmitted from mother to son and never from father to son. Of course, a heterozygous mother transmits the mutant gene to half the daughters, the other half receiving the normal allele. An affected father transmits the mutant gene to all his daughters, since he has only the one X chromosome bearing the mutation.

The most familiar example of X-linked recessive inheritance is hemophilia, the bleeding disease of the Royal House of Hapsburg. More recently, the Lesch-Nyhan syndrome has been identified as another X-linked recessive disorder. It was first described in 1964, when the two pediatricians,[15] whose names it now bears, observed two mentally retarded brothers with self-mutilating behavior, choreoathetosis, and hyperuricemia. The syndrome is now known to result from a deficiency of the enzyme hypoxanthine guanine phosphoribosyl transferase (HGPRT).[16] As an X-linked recessive disorder, it occurs only in males and is transmitted by unaffected carrier mothers to their sons; no affected fathers have been found. Even though some of the affected individuals survive to childbearing age, the severe mental deficiency and self-destructive behavior seemingly are incompatible with procreation. Unlike Huntington chorea and Wilson disease, Lesch-Nyhan syndrome is manifested very early in life and, not being transmitted by affected individuals, presents none of the problems discussed previously. Although the enzymatic defect has been known for six years and its role in purine metabolism explored, there remain many open questions, and at the moment there is no known therapy.

It is possible, however, to diagnose Lesch-Nyhan syndrome in utero, so that in a family with one affected individual, amniocentesis can reveal whether a male fetus is affected and whether a female fetus is a carrier.

X-LINKED DOMINANT: MANIC-DEPRESSIVE DISEASE (?)

X-linked dominant inheritance implies that a single mutant gene on one of the two X chromosomes will be manifested by the carrier despite the presence of the normal allele on the other X chromosome. Consequently, both males and females who carry a single X-linked mutant gene will manifest the trait in question. Until recently, X-linked dominant inheritance in man was known only for the Xg blood group system.

Lately, however, evidence has begun to accumulate that one of the psychoses commonly classified as functional in the textbooks, bipolar manic-depressive illness, may actually result from a single gene defect transmitted in some families according to an X-linked dominant mode of inheritance. As early as 1953, Kallmann [17] stated unequivocally that manic-depressive psychosis was dependent on the mutative effect of a single irregularly dominant gene. Although Kallmann considered incomplete penetrance more likely than X-linkage, he was thoroughly convinced of the dominant mode of inheritance. The data of Winokur and Tanna [18] and of Mendlewicz and associates [19] suggest X-linked dominant inheritance, at least in some families. It is too early to tell whether this will be found to be the most common, or even a common form of transmission. Indeed, because of many problems (father-son transmission,[20] limited data on linkage to both Xg and color blindness alleles, pedigrees spanning only two generations, etc), further data will be needed before this mode of inheritance can be considered even for the few families reported so far.

In X-linked dominant inheritance, the presence of a single mutant gene on one of the two X chromosomes is sufficient to produce the disease. Since women carry two X chromosomes and men only one, X-linked dominant conditions are expected to be more frequent in women than in men. Theoretically, if the dominant gene comes from the father, then all of his daughters will be affected and none of his sons, the sons inheriting their father's Y chromosome, the daughters his X chromosome. For an affected mother there is an equal probability of having affected sons or affected daughters, 50 percent in each case, since each of her children has a 50 percent chance of getting the mutant X chromosome and a 50 percent chance of getting the normal X chromosome.

In actuality, there are many affected sons of affected fathers, and the genetics of manic-depressive psychosis are far more complex than indicated by the sketchy remarks above. It is likely that a number of etiologically distinct disorders will emerge from what is now a single diagnostic category, quite aside from the current distinction between bipolar and unipolar disease. The X-linked mode of inheritance refers only to bipolar disease, even though it is not uncommon for certain members of families in which bipolar disease occurs to manifest symptoms of only unipolar disease. However, it is generally believed that those individuals who show depression only, do so as a result of an altered expressivity of the same bipolar genotype. By contrast, persons with unipolar disease from families showing only unipolar disease are believed to carry a different genetic abnormality than families with bipolar disease. A detailed analysis of information available to date is provided in the chapter by Cadoret and Winokur.

Attempting to delineate homogeneous subgroups of depressive illness is important not only to the geneticist for theoretic purposes but also to the psychiatrist for therapeutic reasons. Thus, for example, it appears that patients with a history of bipolar illness among relatives respond better to the prophylactic use of lithium carbonate than do patients without such a history.[21,22] A strong family history of bipolar disease should, therefore, suggest to the clinician a trial of lithium prophylaxis even if the patient manifests unipolar rather than bipolar disease.

Incidentally, manic-depressive psychosis, like Wilson disease, illustrates the fallacy of the commonly held belief that if a disorder is inherited, it is therefore inevitably impervious to attempts at amelioration. We now have not only effective treatment for this disorder but also the possibility of aborting further attacks of mania and depression once the disease has become manifest.[23,24] Instead of engendering an attitude of hopelessness regarding therapeutic intervention, the demonstration of a genetic predisposition to a given dysfunction should lead to hopefulness. Once the organic mechanisms can be delineated, and sometimes even long before that, rational therapy can be devised.

One of the consequences of effective prophylaxis and therapy may well be an increased reproductive rate, with an augmentation of the frequency of the deleterious gene(s) in our population. That development has already taken place in the case of schizophrenia, where the acceleration in reproductive rates following the introduction of psychotropic drugs has exceeded that of the population at large.[25]

UNDETERMINED MODE OF TRANSMISSION: SCHIZOPHRENIA

In schizophrenia, the most common of all psychiatric disorders, the exact mode of transmission has not yet been established. It is not for lack of studies or for lack of cases, but for want of agreement in the interpretation of available data that the inheritance of schizophrenia remains an open question. Genetic models have been proposed that involve a single major gene, recessive or dominant with incomplete penetrance, with or without modifier genes [17,26–33]; two genes [34–36]; many genes (polygenic) [37–40]; or more than one mode of transmission (heterogeneity).[41,42]

In schizophrenia it appears that, even with the most sophisticated statistical techniques, it is not possible to distinguish between single-gene and polygenic models. Kidd and Cavalli-Sforza, for example, conclude on the basis of the analysis of a threshold model that both polygenic and single-gene hypotheses are in approximate agreement with the data.[43] They suggest that the solution is unlikely to come from statistical techniques alone, but will have to await physiologic, toxicologic, or biochemical advances.

Meanwhile, the following concepts have been of prime importance in genetic investigations.

Twin studies. The rationale underlying twin studies is that for any trait with a significant hereditary component, one-egg (monozygotic, uniovular, or identical) twins will be significantly more alike than two-egg (dizygotic, binovular, or fraternal) twins. Both types of twins share relatively similar environmental circumstances, being born of the same mother after having shared a similar intrauterine environment, and being brought up in the same family under the same socioeconomic circumstances and in the same cultural milieu. The major difference between the two types of twins is that one-egg twins have identical genotypes (except for spontaneous mutations or chromosomal changes), whereas two-egg twins are no more alike genetically than ordinary sibs born at different times, although their environments are probably more similar. Accordingly, significantly greater similarity (higher concordance) in one-egg twin partners than in two-egg twins in the development of any disorder is taken as evidence of important genetic determinants of that disorder.

A frequently voiced objection to the twin study method is that there may be factors in the environment, as yet unidentified, which are crucial to the development of a disorder and which are present in varying degrees in the families of affected patients. In order to disentangle the effects of heredity and environment, twins separated early in life and brought up apart are most useful. Unfortunately, only few such pairs have been described.

Family studies. The most complete data on the frequency of schizophrenia among relatives of schizophrenic index cases were provided by Kallmann.[31] He found that the schizophrenia risk increases in direct proportion to the increase in degree of genetic relationship. Thus, the lowest risk is faced by step-sibs (no genetic relationship to the index case) and the highest risk by monozygotic co-twins (completely similar with regard to genetic endowment). The risk for dizygotic co-twins is closely similar to that for sibs, as would be expected on the basis of the genetic hypothesis, since sibs and dizygotic co-twins share the same proportion of genetic material, i.e., 50 percent on the average. Half-sibs, with only half the genetic commonality of full sibs, showed lower risks.

Adoption studies. The study of adoptees is the reverse of, and complementary to, the study of identical twins reared apart. In separated identical twins it is the effect of varying environments upon the same genotype that is subject to investigation. In particular, the twin reared by the natural parents (where biologic and environmental influences are compounded) is compared with the twin reared by persons to whom he has no biologic relationship, so that biologic and environmental influences are separated. In the study of adoptees, individuals are reared by persons to whom they have no biologic relationship (adoptive parents).

If biologic factors are important in the etiology of a given trait, then a higher frequency of the trait will be found among natural parents than among adoptive (rearing) parents. If child-rearing practices exert a significant influence on the development of a given trait, then a higher frequency of that trait would be expected among rearing parents than among natural parents.

Another useful approach consists in comparing adopted children with natural children in the same household. Here, persons who have no genetic relationship to one another (natural and adopted children) are reared in the same home environment, so that biologic and environmental factors can again be separated. This approach is being used in studies of intelligence.[44]

Because of the careful safeguards erected by adopting agencies to protect the anonymity of the adoptee, adoption studies were considered impossible in this country until Heston demonstrated their feasibility.[45] Heston gathered his experimental subjects from children born to schizophrenic mothers in a psychiatric hospital and separated from them within 3 days of birth some 20 to 30 years earlier. He found 47 such children and selected 50 control subjects from foundling homes, obtaining the control subjects by alternating between admissions prior to and subsequent to that of the index case, until the appropriate child could be found. These control children were born to normal parents and were matched for sex, type of eventual placement, and length of time in child care institutions. When these offspring were evaluated, it turned out that five of those born to schizophrenic mothers were diagnosed as schizophrenic, but none of the offspring of the control mothers. This difference is statistically significant.

A number of adoptive studies followed. They are discussed, together with the results of twin studies, in the chapter by Liston and Jarvik.

Penetrance and expressivity. All three types of studies discussed above—twin studies, family studies, and adoptee studies—point toward hereditary determinants in the etiology of schizophrenia. Nonetheless, environmental factors must also exert a significant influence on the development of this disease, inasmuch as monozygotic twins, whose genotypes are identical, do not show 100 percent concordance. At the moment, we are unable to identify those environmental factors that are crucial either in precipitating a schizophrenic episode or in protecting a vulnerable individual from a schizophrenic reaction. As pointed out earlier, regardless of the mode of transmission (which remains to be established for schizophrenia), we know that monozygotic co-twins of schizophrenic patients carry the genetic information necessary for the development of the disease. The fact that a certain proportion of these co-twins fails to manifest the disease is an indication of reduced penetrance. Penetrance is generally defined as the percentage of individuals with the appropriate genotype who actually manifest the given condition. Huntington chorea is believed to be fully penetrant—that is, 100 percent of genetically susceptible individuals develop the disease provided they survive the manifestation period. Penetrance in schizophrenia is still unknown but is probably greater than 50 percent. Even though the average concordance rate for monozygotic twins, gleaned from the world literature, is only about 50 percent, many of the discordant pairs later become concordant. Long-term follow-ups, from which an exact percentage could be derived, have not yet been carried out. Since the mode of inheritance is as yet undetermined, parental genotypes cannot be identified, and our answer will have to come from long-range studies of identical twins. It may be necessary to extend such studies beyond the age usually designated as the end of the manifestation period (45 years), since psychotic disturbances in later life, such as the late paraphrenias and some senile psychoses as described by Kallmann [46] and Roth,[47] may represent formes frustes of the same genotype.

In the face of the very limited information gathered to date, it would seem that highest priority should be given to the task of identifying those environmental factors that have protective value for the schizophrenia-prone genotype. Even if such information would not lead to complete prevention of the disease in a significant number of vulnerable persons (reduction in penetrance), it would lead to a marked delay in onset and a milder course, i.e., reduced expressivity.

In contrast to penetrance, which involves the presence or absence of a given condition in a vulnerable genotype, expressivity denotes clinical course and severity in affected individuals. Were we in a position to alter the expressivity of schizophrenia through appropriate manipulation of gene-environment interaction, we would have made significant progress.

The frequency (per 100) of these gene-borne abnormalities in the general population is:

Huntington chorea	0.004–0.007
Wilson disease	0.001
Lesch-Nyhan	extremely rare *
Manic-depressive	0.3–0.4
Schizophrenia	0.86

* No data have yet been published.

Chromosomal Abnormalities

Unlike the gene-borne inheritance, which cannot be visualized in man and in which the location of the abnormal gene may be anywhere on the 22 pairs of autosomes (unless the trait is X-linked), chromosomal abnormalities are microscopically visible and, therefore, provide incontrovertible evidence of a change in the hereditary material. Human chromosomes have been adequately studied for only two decades, i.e., after the introduction of tissue culture as a source of rapidly dividing cells and the discovery of new techniques for harvesting and fixing cells. The normal chromosome complement (diploid number) has been established as 46 (22 pairs of autosomes and one pair of sex chromosomes). The chromosome constitutions (karyotypes) of men and women differ in only a single pair, the sex chromosomes. Whereas normal women have two X chromosomes, normal men have one X and one Y chromosome (Figure 3); the same Y chromosome is passed—usually along with the family name—from father to son through successive generations. Occasionally, there are cells with more or less than 46 chromosomes, but normally these hyper- or hypodiploid cells are seen in less than 5 percent of cells examined (except for the aged, where the frequency is higher), and different cells show loss or gain of different chromosomes. In cases where there is a specific, rather than a random, gain or loss of chromosomes, affecting a significant proportion of cells, we speak of *mosaicism,* i.e., the presence within a single individual of two or more cell lines, derived from a single stem line. There are many examples of mosaicism in sex chromosome abnormalities, e.g., XO/XX/XXX, where in a single person all these cell lines are observed. In one case of sex chromosome mosaicism, the respective frequencies were 60 percent XO, 26.7 percent XX, and 13.3 percent XXX.[48]

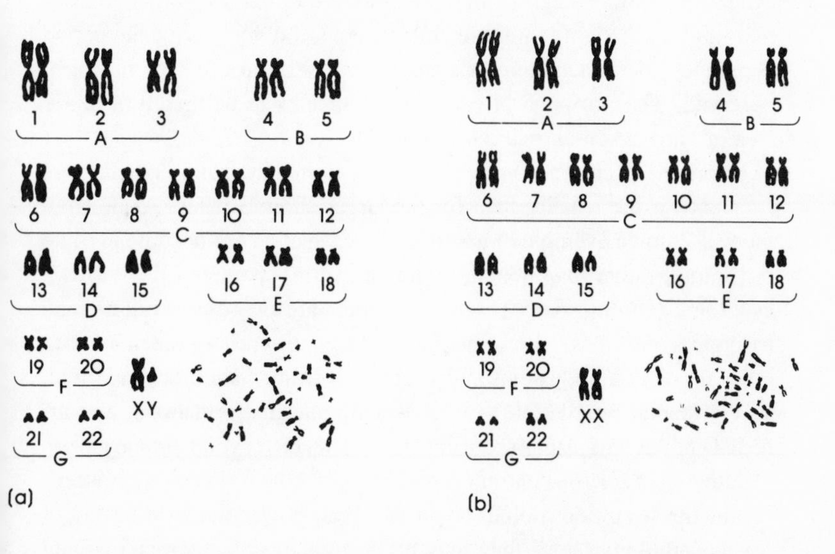

Figure 3. Karyotypes of (a) a normal male and (b) a normal female.

Structural chromosome abnormalities, as distinguished from numerical ones, also occur with a relatively high frequency. These include breaks, translocations (breaks in two chromosomes, followed by transfer of a segment from one chromosome onto the other), deletions (loss of a portion of a chromosome), and ring chromosomes (deletion of distal ends of both arms, followed by fusion of the deleted ends to form a ringlike configuration).

It should be pointed out that chromosomal abnormalities do not affect every one of the 23 pairs of chromosomes equally. Indeed, the commonly occurring chromosomal abnormalities are limited to relatively few chromosomes. Moreover, except for the X chromosome, absence of an entire chromosome has not been verified in a living human being and may not be compatible with survival. The absence of a less than complete chromosome is best exemplified by the partial deletion of a B-group chromosome, which gives rise to a clinical syndrome distinguished by a high-pitched cry (hence the name cri-du-chat, or cat's cry syndrome). Children with this syndrome are retarded; those initially described were all severely retarded,[49] but with increased attention to the syndrome, milder degrees of retardation have been recorded. Originally, it was thought that these children, too, did not survive the period of infancy, but since then considerable variation in life span has been observed. The profound effect of what appears to be merely a minor loss of chromosomal material is sobering.

Excess of a chromosome is relatively common, but is much more frequent for the smaller than for the larger chromosomes. Aside from the well-known example of trisomy (presence of a chromosome in triple or triploid, rather than double or diploid, form) characteristic of Down syndrome (chromosome 21 trisomy), there are two other well-defined trisomic syndromes, one affecting the D-group chromosomes and another the E-group chromosomes. Both syndromes are associated with severe mental retardation as well as anatomic malformations, and affected individuals usually do not survive much beyond infancy.[50–53]

Although numerous chromosomal abnormalities are known, those affecting the sex chromosomes are of the greatest interest to the student of psychopathology, since they may occur without accompanying mental deficiency and may have a profound influence on psychosexual development. These disorders are particularly important for genetic counseling because affected individuals are often entirely asymptomatic. In the case of infants with Down syndrome, the diagnosis can usually be made on purely clinical grounds, and recourse to cytogenetic techniques is required only for detection of the specific abnormality (trisomy, translocation, mosaicism). Extra sex chromosomes in the newborn, however, are generally discovered unexpectedly as the result of routine screening procedures.

Abnormalities of sex chromosomes lend themselves to research much more readily than those of any other chromosomes because a lack or excess can be detected without recourse to cell cultures, a laborious procedure at best. It is possible to take a sample of tissue (the most popular is a scraping of the buccal mucosa) to fix and stain, and then examine the slides for a dark staining chromatin body at the periphery of the nuclear membrane. This chromatin (or Barr) body is seen in about a third of the female cells, whereas in males it is not seen at all. The chromatin body represents the second X chromosome, only one of the two X chromosomes being active in any female cell. Normal males, having only a single X chromosome, have no inactive X chromosome and, therefore, do not have a chromatin body. In the case of chromosomal abnormalities, the presence of extra X chromosomes is signaled by the presence of extra chromatin bodies, there always being one less chromatin body than the number of X chromosomes. This technique, unlike karyotypic examinations, is neither time-consuming nor demanding of skilled technicians. Analogous to the ordinary differential blood count, the technique lends itself readily to mass screening, for which it has been widely used.

Recently a technique has been developed for the detection of the Y chromosome. For this purpose, special fluorescent staining methods are required; once these have been worked out, they are not much more demanding than the techniques used to detect the chromatin bodies in females, although it is not certain at this time whether they are as reliable. It must be stressed that the fluorescent techniques pick up the presence of a single Y chromosome, as well as extra Y chromosomes. In those cases where more than one Y chromosome is present, there is also more than one Y body, so that the number of Y bodies corresponds to the number of Y chromosomes in the cell. Female cells have no Y chromosome and, therefore, no Y body; normal male cells with a single Y chromosome have a single Y body.

MISSING SEX CHROMOSOME: TURNER SYNDROME (45,XO)

Turner syndrome was known to medicine long before there was any conception of chromosomal abnormalities in man. It encompasses a widely varied symptomatology accompanying the lack of a second sex chromosome. Those girls in whom it occurs have but one X chromosome, no Y chromosome, and no Barr body. The only consistent clinical feature is congenital ovarian deficiency with sexual infantilism and primary amenorrhea, accompanied by high gonadotropin and low estrogen levels. It is characterized by "streak ovaries" containing rete ovarii and hilus cells as well as ovarian stroma, but no follicles.

Other abnormalities, often part of the syndrome, are dwarfism, webbing of the neck, low hairline, shield chest, multiple pigmented nevi, and skeletal and cardiac anomalies, particularly coarctation of the aorta. The majority of these are contained in Turner's description of the syndrome.[54] There are numerous other somatic manifestations, and the syndrome can in some cases be diagnosed at birth by the presence of marked edema of the limbs, as pointed out in the early descriptions by Ullrich.[55] The list of abnormalities that are sometimes present, but certainly not always, covers more than 2½ pages in a recent text.[56] So varied are the clinical manifestations that in excess of two dozen synonyms exist to describe the condition,[57] and attempts are being made to construct separate subcategories.

This clinical variety is almost matched by the diversity of chromosomal abnormalities, which number at least 15,[58] and range from the 45, XO chromosome constitution described above, through 46, XX and 46, XY, to the various deletions of one of the two X chromosomes, structural abnormalities of one X chromosome, and a plethora of mosaics. In the absence of a Y chromosome, secondary sex characteristics are female, despite the fact that only one, rather than two, X chromosomes is present. Incidentally, this tendency toward female sexual development in the absence of a male-determining chromosome is characteristic of all mammalian species. Even though most of the phenotypes are female, there are also "male Turners," and there are 46, XY "females" as well as 46, XY male pseudohermaphrodites.

Considering its heterogeneity, it is not surprising that the incidence of the disorder is not too well established. Since these patients classically have but one X chromosome, they are chromatin-negative (no chromatin body), and the frequency of chromatin-negative cases ranges from 1/1000 to 1/3000 newborns. The frequency is not greater either in institutions for mental defectives or in institutions for the mentally ill, and does not rise with parental age. Since patients with gonadal dysgenesis are sterile, there need be no concern about direct transmission of the disorder to future generations. However, even though Turner syndrome is not "hereditary" in the usual sense of the word, there may be, in some families, a predisposition toward the development of this anomaly, as indicated by certain pedigrees.[58]

Another puzzling aspect of the 45, XO anomaly is that while it is so lethal prenatally that only 1 out of 40 zygotes may come to term,[59] for most of those born alive it is compatible with survival into adulthood. There is as yet no resolution of this apparent contradiction.

The question of intellectual impairment is also a moot one. Although the initial reports pointed out that these girls were not retarded, subsequent publications emphasized a slight degree of intellectual deficit, particularly in girls with marked physical deformities. Mental retardation, however, is not characteristic of Turner syndrome, and all levels of intelligence are represented. There does seem to be a more or less specific impairment, termed "spatial agnosia" by Money, which includes difficulty in right-left orientation, and there is also relatively poorer arithmetic performance than verbal performance.[60] It has also been proposed that there may be a general performance deficit, as distinct from a specific spatial agnosia.[61] It is of interest that the abilities most impaired in Turner syndrome are those generally more highly developed in males than in females. The hierarchy of performance goes from normal male to normal female to Turner syndrome—normal males doing best, and patients with Turner syndrome worst, with normal females in between. During recent years, a number of investigators have attempted to delineate the hereditary component of spatial ability: the data all point toward sex-linked factors.

The situation is still far from clear, but the best available evidence indicates the involvement of the sex chromosome in the transmission of spatial ability. Stafford hypothesized a sex-linked recessive mode of inheritance of spatial ability on the basis of two observations.[62] First, males had a higher mean score on spatial tasks than females, an observation known for some time. Second, a unique pattern of correlations arose from Stafford's data: unlike-sex parent-child pairs (i.e., mother-son and father-daughter) exhibited a higher correlation on spatial test scores than like-sex parent-child pairs (i.e., mother-daughter and father-son).[61] These observations imply a sex-linked recessive inheritance but are not inconsistent with autosomal inheritance. Support for sex linkage also comes from McClearn,[63] and familial correlations similar to Stafford's were seen in the work of Hartlage.[64] Finally, with a larger sample population, Bock and Kolakowski arrived at similar correlation patterns; in combining the results of these three studies, they provided convincing evidence for X-linked recessive inheritance for spatial ability.[65]

Of particular interest to psychiatrists is the psychosexual development of a group of individuals essentially without gonads. First, neither psychiatric disorders nor homosexuality appear to be any more common in this group than in the population at large.[66] Since most of the patients with Turner syndrome are phenotypically female, they are generally brought up as girls. Money, who with his associates has carried on the most extensive studies in this area, concludes that these girls are unequivocally feminine, with age- and sex-appropriate play, daydreams, interests, and attitudes. During adolescence, however, they tend to be greatly concerned about their physical appearance, and the development of emotional maturity seems to be directly related to the rate of physical maturation. Money found that those girls most delayed in physical development and most dwarfed tended also to be the most childish and conforming. In general, girls with gonadal dysgenesis grow up being fond of children and make excellent adoptive mothers.

EXTRA SEX CHROMOSOME

Triple-X females (*47,xxx*). In a sense, the 47,XXX genotype is just the opposite of the 45, XO genotype, there being an extra X chromosome rather than a missing one. These women are generally free of any major congenital malformations. They exhibit no physical stigmata, have normal menses, and are fertile. They do, however, show a higher frequency of mental retardation than the population at large. While the frequency of XXX among newborns (about 1/1,000) is on a similar order of magnitude as the frequency of XO, in institutions for mental defectives it is higher, ranging from 2.6/1,000 to 6.7/1,000.[56]

The frequency of XXX females at birth has been reported as high as 0.12 percent [67] and is considerably higher in institutionalized mental retardates. In addition to the higher frequency of mental retardation, the triple-X females were also said to be more frequent in the offspring of older parents (more recent data do not confirm this). Studies of the Xg blood groups suggest that usually the extra X chromosome is of maternal origin.

Unlike patients with Turner syndrome, XXX females are not sterile and are known to have reproduced. By 1972 the literature contained information on at least 13 triple-X women who produced 23 boys and 12 girls, all of whom were phenotypically normal and had normal sex chromatin. This finding is unexpected, since upon meiosis one would assume that half the gametes would contain a single X chromosome and half two X chromosomes, so that half of the offspring would have an extra sex chromosome (either XXX or XXY). Hamerton suggests as an explanation "some form of directive segregation during oogenesis so that the disomic chromosome complement is regularly included in one of the polar bodies" (p. 106 [58]).

There is an increased frequency of the XXX karyotype not only in institutions for mental retardates but also in institutions for the mentally ill (the major psychiatric diagnosis being schizophrenia or paraphrenia). The increase in both mental retardation and mental illness may represent a nonspecific effect of an excess of chromosomal material. With regard to mental retardation, the higher the number of excess X chromosomes, the greater the frequency of intellectual impairment and the more severe its degree, be it in males or in females. Patients with four or more X chromosomes, for example, are almost invariably retarded. This is also true for extra Y chromosomes, and all of the autosomal trisomies are associated with mental retardation. In those chromosomal disorders not inevitably accompanied by severe mental deficiency (i.e., sex chromosome abnormalities), behavioral disturbances have come to the fore to a greater extent than in the general population. Thus, there is a higher frequency than expected in psychiatric hospitals not only of XXX but also of XXY and XYY. As was mentioned under Turner syndrome, the same does not hold true in cases where there is a deficiency of chromosomal material.

In view of the singular lack of physical abnormalities in XXX females, the increased frequency of psychosis is difficult to explain as a reaction to the chromosomal abnormality itself, an explanation adduced to account for the psychiatric disturbances exhibited by persons with other extra X- or extra Y-chromosomal disorders. The XXX abnormality is usually unknown to the individual who exhibits it, until it is discovered accidentally in a chromosomal survey.

A comparison of XO, XX, and XXX individuals should provide us with an idea of the function of the X chromosome, and to some extent it has. Thus, we know that a single X chromosome is not sufficient for gonadal development, and that an excess of X chromosomes (three or more) does not interfere with normal sexual maturation and fertility. Furthermore, women with a single X chromosome are usually short, and those with extra X chromosomes tend to be taller than average. Factors influencing growth are apparently located on the short arm of the X chromosome.[68]

When it comes to intellectual functioning, we know that average and even superior cognitive abilities can exist despite the lack or excess of an X chromosome. The presence of only a single X chromosome appears to be accompanied by specific intellectual deficits, but it is still undetermined whether an extra X chromosome has specific cognitive effects. Extensive psychologic testing primarily designed to detect specific intellectual changes remains to be done, since this question has been largely ignored in the work-up or write-up of the XXX cases reported to date (there are over 60 of them). However, most of those cases were discovered in institutions for the mentally retarded; relatively few individuals with average intelligence were included. It is the latter who could provide the most interesting data, since whatever specific deficits might be associated with an extra X chromosome would not be obscured by generalized mental retardation and a concomitantly low level of intellectual performance.

XXX girls also offer an opportunity to assess the influence, if any, of an extra X chromosome upon psychosexual development. Here, too, those cases with normal or better intelligence would be most valuable.

Klinefelter syndrome (47,xxy). The male equivalent of the XXX syndrome is Klinefelter syndrome, first described in 1942, more than a decade before the chromosomal basis of the disorder became known.[69] Patients with Klinefelter syndrome in its simplest form have an extra X chromosome with an XXY chromosomal constitution, but the eponym "Klinefelter syndrome," like "Turner syndrome," subsumes a large number of variants. Up to five X chromosomes have been reported (XXXXXY), as well as XXYY, other combinations of extra X and extra Y, and numerous mosaics. The presence of the Y chromosome leads to phenotypic maleness in Klinefelter syndrome, regardless of the number of X chromosomes.

The hallmarks of the syndrome are seminiferous tubular dysgenesis with sterility and gynecomastia. Although the symptomatology varies, the characteristic signs are small testes with hyalinized seminiferous tubules and absence of spermatogenesis in the presence of intact Leydig cells. The existence of small testes is usually not noted until puberty, when the expected enlargements fail to occur. Gynecomastia is not always present, nor are feminizing or eunuchoid features always observed.

There is an increased frequency of mental retardation. Whereas Klinefelter occurs approximately once in 500 newborn males, it has been found as often as once in 100 in institutions for mental retardates. Again, the greater the excess of chromosomal material, the greater the frequency of mental deficiency and the more severe its degree. For example, in the variant of Klinefelter syndrome with three rather than one extra X chromosome, the range of IQ for the 29 cases in which IQ was recorded (out of 70 such cases in the literature) was from 19 to 73, with only a single individual having an IQ above 59.[70] The mean IQ for this group was 38.2. Among the 48 for whom no IQ data were provided, there may be a single exception, a 15-month-old child said to have normal intellectual development.

Although statistically XXY patients show a marked increase in the frequency of mental retardation, most are probably not retarded, and some may even be of superior intelligence; some are college and medical school graduates. Patients with Klinefelter syndrome most often come to medical attention either at puberty, because of gynecomastia, or later in life, because of infertility; they constitute between 10 and 20 percent of the patient population at infertility clinics. There is a wide range of symptomatology. Consider gynecomastia—in some persons it is not present at all, in others it is so mild as to be barely perceptible, and in still others it is so obvious that it creates considerable psychic trauma and frequently leads to plastic surgery. Many of these males have serious problems in psychosexual development, even in the absence of gross malformations; following puberty, unless treated, they suffer from testicular atrophy and all the symptoms resulting therefrom. The entire problem will be discussed in depth in Stoller's chapter and is introduced here merely for purposes of orientation.

Like XXX individuals, persons with Klinefelter syndrome are found more frequently than expected in psychiatric hospitals, often with the diagnosis of schizophrenia. This type of psychopathology, in contrast to the psychosexual problems, may be a nonspecific effect of the extra chromosomal material. Further, Klinefelter patients frequently have a history of impulsive, antisocial, and criminal behavior but, as will be discussed below, not nearly to the same extent as persons with an extra Y chromosome. Since, as far as is known today, women with an extra X chromosome do not exhibit socially deviant behavior, it is unlikely that the X chromosome is implicated in such behavior. It is not entirely inconceivable, however, that an imbalance between X and Y chromosomes may play a role.

In line with the evidence for growth-promoting factors on the short arm of the X chromosome, patients with Klinefelter syndrome tend to be taller than average.

Since patients with Klinefelter syndrome are sterile (the occasional exceptions reported have been interpreted as due to undetected mosaicism), the anomaly is again not transmitted from one generation to the next. Although there are some reports of a predisposition to nondisjunction, including Klinefelter syndrome, and even though the association between Klinefelter syndrome and Down syndrome is higher than would be expected by chance alone, other studies conclude that no familial predisposition toward nondisjunction has been demonstrated.[58]

Like the other sex-chromosome abnormalities, Klinefelter can be diagnosed in utero, or at any time after birth, so that it may be picked up in routine screening procedures.

Extra Y Syndrome (47,xyy). The final example is also the most controversial. At a time when hundreds if not thousands of cases with other chromosomal abnormalities had been recorded, the world literature contained less than a dozen isolated case histories of individuals with an extra Y chromosome. Then there appeared an article reporting a high frequency of the extra Y chromosome in an institutition for mentally retarded prisoners requiring maximum security.[71] In that group, if only persons over 6 feet tall were considered, the frequency of the extra Y chromosome was 50 percent. Outstanding features of the extra Y syndrome were great height, aggressive behavior, and borderline intelligence. Otherwise, they had no physical stigmata that would differentiate them from other males, patients or not. The same had not been true for those cases reported earlier in the literature, many of whom had shown urogenital and other abnormalities. Here, then, for the first time, a specific behavioral abnormality other than cognitive deficit was linked to a chromosomal aberration.

Soon other investigators began to look for persons with an extra Y chromosome among tall, retarded criminal offenders, and within a few years over 100 cases had been reported in print. Because screening techniques for an extra Y chromosome were not available at the time, relatively few cases were examined, and most investigators restricted themselves to those population groups where the likelihood of finding persons with the extra Y chromosome was highest; i.e., they looked among retarded, tall, violent individuals. Clearly, such a research design is not an optimal one, and it is conceivable that an adequate study of entire population groups might reveal that individuals with the extra Y chromosome are as frequent in populations of average height, intelligence, and aggressiveness as they are among tall, violent criminals. Indeed, reports began to appear of individuals with an extra Y chromosome who did not show either aggressive or criminal behavior—notably one of the early cases, a kindly family man whose only reason for coming to the attention of the researchers was that he volunteered to donate a blood sample, which was then used for chromosome studies. As a result of such cases, a number of articles began to appear in the lay press, the medical newspapers, and the scientific literature itself, proclaiming that men with an extra Y chromosome had been unjustly maligned and that the association between violent behavior and the chromosomal abnormality was a myth created and perpetuated by the mass media.

On the other hand, nonmammalian studies have demonstrated that in the Japanese medaka, a species of fish in which it is possible to breed individuals with a YY chromosome constitution, the YY fish were longer, more aggressive, and more successful in the chase, fertilizing by far the largest proportion of eggs in the tank they shared with normal XY fish.[72] But it behooves us to be leery when transposing findings from fish to man.

Fortunately, by now, nine years after the first implication of the extra Y chromosome in aggressive behavior, there exist sufficient data on man to warrant an examination. We recently completed such a review,[73] and the findings from the world literature are summarized in Figure 4. In order to provide some anchor for the data on XYY, it was deemed imperative to have some comparison group. Fortunately, many of the studies were based on complete examinations of karyotypes, so that abnormalities other than extra Y chromosomes were also reported. Since an extra X chromosome is by far the most common chromosomal abnormality in adult males, we decided to include only those studies where examinations included a search for both X- and Y-chromosome abnormalities. The advantage of utilizing the XXY chromosomal constitution for comparative purposes rests on two facts. First, some authors maintain that criminal aggressive behavior is as frequent in patients with Klinefelter syndrome as it is in persons with an extra Y chromosome. Second, the chromatin-body technique, unlike cell culture and karyotypic examination, lends itself to large-scale screening. A vast number of additional cases with just the chromatin-body data are therefore available for comparative purposes. As can be readily seen from Figure 4, in eight studies of newborn males from Scotland, Canada, and the United States, a total of 17 individuals with an extra Y chromosome were detected, for a frequency of 0.14 percent, and 18 individuals with an extra X chromosome, for a frequency of 0.15 percent.

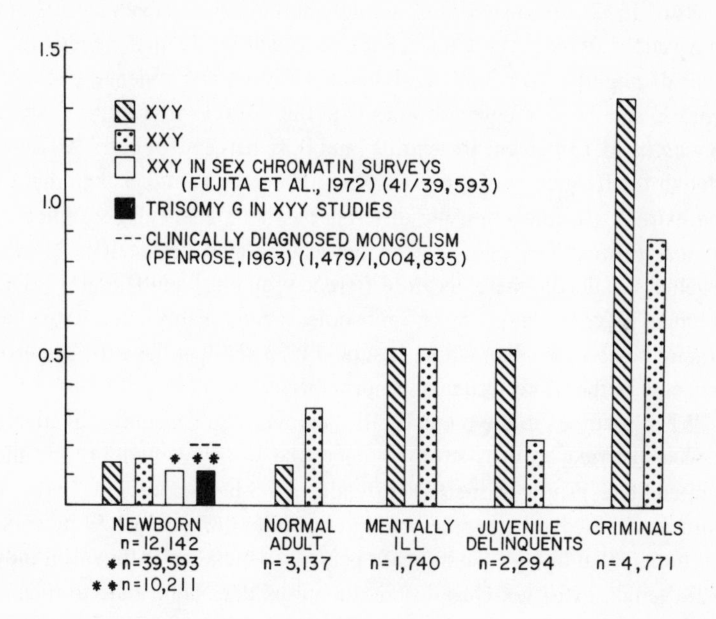

Figure 4. Comparative frequencies of XYY and XXY males (studies designed to detect both).

During the same period, there were eight studies in Great Britain, Switzerland, the United States, the Congo, and Japan, with a total in excess of 39,000 individuals on whom surveys of sex chromatin were carried out. Among them the frequency of extra X chromosomes was 0.11 percent. The correspondence between 0.11 percent and 0.15 percent is sufficiently close to allay any doubts regarding the acceptability of the data on extra X chromosomes and an extra Y chromosome in the much smaller samples with karyotypic analysis. Interestingly enough, the difference between clinically detected Down syndrome and chromosomally verified diagnosis is similar to that observed in cases of the extra Y chromosome. The clinical data, as summarized by Penrose[75] for over 1 million infants, provided a rate of 0.15 percent, while 0.11 percent was the average based on 10,211 cases with chromosome analyses.[73,76,77]

As of 1972, seven studies of normal adult males had been carried out in Sweden, France, the United Kingdom, and the United States, on a total of slightly over 3000 individuals. The correspondence between extra X and extra Y chromosomes is not as close as it is in the case of newborns: 0.13 percent for extra Y and 0.32 percent for extra X. Even though the frequency of the extra X chromosome is higher than that of the extra Y chromosome, the difference is not statistically significant, so that further data will have to be accumulated before it is known whether, indeed, there is a difference among adult males. The frequency for the extra Y chromosome is remarkably similar to that among newborns. Meanwhile, it should be noted that the extra X chromosome is the more frequent abnormality.

When it comes to the mentally ill, however, the six studies available to date (carried on in Norway, France, the United Kingdom, and the United States) show an increased frequency for both extra X and extra Y chromosomes, the percentage being uncannily the same (0.52 percent) for both. Even though the increase is not statistically significant, it may well be that extra sex-chromosomal material does contribute to mental illness, particularly schizophreniform syndromes. Many of the individuals with extra Y and X chromosomes do show such mental disturbances, and there is a high frequency of psychopathology among XXX.

It should be noted at this point that in all three groups discussed so far—newborn males, normal adult males, and mentally ill males—the frequency of extra Y chromosomes is no greater than that of extra X chromosomes. Indeed, the two seem to be closely similar, with the extra X more frequent. The situation is different when it comes to populations of delinquents and criminals, as shown by the seven studies of juvenile delinquents reported from the United Kingdom and the United States, which yield a frequency of 0.52 percent for extra Y chromosomes and 0.22 percent for extra X chromosomes among somewhat over 2,000 individuals. This difference, though more than twofold, is not statistically significant. That is not the case with adult criminals, however. Over 4,000 persons were examined in 23 studies carried out in Denmark, France, Ireland, Australia, the United Kingdom, the United States, and Mexico. The difference is smaller—the frequencies for extra Y and extra X chromosomes are 1.34 percent and 0.88 percent, respectively—but it is significant at the 5 percent level of confidence.

It bears emphasis, however, that the violent behavior is episodic and that XYY men are not continually aggressive persons.[78] In general, the impression gleaned from the early studies and from individual case histories has been validated; i.e., violent behavior is associated with an extra Y chromosome to a greater extent than it is with an extra X chromosome. The same holds true for increased stature. Clearly, not every person who has an extra Y chromosome need display aggressive, violent, or antisocial behavior; nor is excessive height a sine qua non of the syndrome; and neither average nor superior intelligence is incompatible with the presence of an extra Y chromosome. Nonetheless, there is sufficient association for the triad so that it must be considered meaningful. Hook, who also surveyed the world literature, reached essentially the same conclusion, even though he divided the cases in a somewhat different fashion.[79] Yet the idea that the presence of an extra Y chromosome could have any untoward effects—aggressive behavior or other features—continues to be questioned. With regard to mental deficiency, for example, it is interesting to observe that few if any investigators ever questioned the association betwen mental retardation and extra X chromosomes, even though the majority of Klinefelter patients are not mentally retarded, and several are known to be of superior intelligence. Likewise, the wide variety of clinical syndromes subsumed under the broad category of Turner syndrome, or gonadal dysgenesis, has been attributed to chromosomal abnormalities, despite the fact that the clinical manifestations are enormously different from person to person, and no clear-cut association has been established between the type of chromosomal abnormality and the symptomatology. Yet when it comes to the presence of an extra Y chromosome, the association with mental retardation is constantly questioned, even though many persons with the extra Y chromosome are of borderline or lower intelligence.

An example from a recent publication will illustrate one of the attitudes:

We suspect that, like many persons of limited intellect, adult XYY males will more often be found at large, at peace with their surroundings in rural areas, than in more stressful urban environs. As the attendant physical features of the

XYY syndrome, such as acne, essential tremor, and excessive height are made the objects of systematic surveys, avoiding bias with regard to intellectual and behavioral traits, a valid composite of the behavioral and intellectual status of the XYY male might soon emerge.[80]

It is beyond the scope of this chapter to examine carefully the reasons for the reluctance to accept untoward intellectual and behavioral effects of the extra Y chromosome. Conceivably, the fact that the Y chromosome is a male-determining chromosome arouses the reaction and accounts for the fact that, insofar as is known, the psychiatric literature contains no mention of the readiness with which detrimental effects are attributed to the X but not to the Y chromosome. Incidentally, none of the studies reported here is based on surveys for Y bodies utilizing a fluorescent technique. Once those data become available, we may have conclusive answers as to the frequency of the extra Y chromosome among normal males, compared with groups of males with varied behavioral abnormalities.

While awaiting such data, what can we predict about the future development of an infant with an extra Y chromosome? We can tell his parents, with a high degree of confidence, that he will probably be quite tall. What can we say about his intelligence? No prediction can be made in any individual case. Given 100 children with an extra Y chromosome, we can predict that the average intelligence will be below normal. We need information on the parents and sibs of persons with the extra Y chromosome in order to compare their intellectual capacities. Even were such information available, we should carefully consider the impact of imparting it to the parents, lest we initiate a self-fulfilling prophecy.

A similar situation exists with regard to the probability of violent, aggressive, antisocial behavior. Unquestionably, such an association exists. Again, however, it exists for groups of individuals but certainly is not present in every individual case. There are many instances where behavior falls in the normal range, and this need not really cause surprise. After all, men with a single Y chromosome show considerable variation in the degree of aggressiveness they manifest. If a man's Y chromosome carries genes that predispose him toward a high level of aggression, and if the man has received two such Y chromosomes, then we can expect the unusually aggressive behavior that leads to criminality. If, on the contrary, another man with a Y chromosome predisposing him toward a much milder temperament receives a second such Y chromosome, then the presence of two such Y chromosomes may merely bring his unusually mild-mannered Caspar Milquetoast behavior into the normal masculine range.

The following statement may be superfluous but still bears mention: genes do not directly influence behavior, and the metabolic pathways and mechanisms of action by which genes exert their influence upon behavior are as yet unknown. In all likelihood, hormonal influences, such as androgens, will be found to be important, and so will a variety of environmental influences that operate in the pathogenesis from gene to behavior and modify basic predispositions.

What is the advantage of identifying persons with an extra Y chromosome? Can we prevent transmission to future generations? Apparently, the extra Y chromosome is not transmitted from father to son. Even though most of the men with an extra Y chromosome have no known progeny, there are a sufficient number who have had offspring, some of them as many as ten; except for a single report, there is no recorded instance of transmission of the extra Y chromosome. Indeed, observation of the spermatozoa demonstrated a diminished efficiency of 24,XYY sperm relative to normal 23,XY sperm in achieving fertilization.[81] Neither is there any reason for concern that future children have an increased risk of this disorder. Further, neither a maternal age effect[82] nor a paternal age effect[83] has been found for the XYY syndrome.

The advantage in following infants for whom the diagnosis of an extra Y chromosome has been made lies in the fact that those who will show unusually aggressive behavior in infancy can be expected to persist in this pattern throughout life unless successful intervention occurs. It is this group that is valuable for testing techniques of interfering with violent behavior, such as minimizing frustration, providing other outlets for aggressive impulses, and administering biochemical and pharmacologic agents that may curb the aggressive impulses and reduce them to a level compatible with normal social intercourse. Indeed, by observing these children, who constitute but a minute fraction of those who will subsequently emerge as criminals or antisocial and violent individuals, we may be able to devise and test techniques that will have wide applicability to all those individuals who exhibit uncontrollable violent behavior without having an extra Y chromosome.

Summary

The purpose of this chapter has been to provide an overview of the types of genetic abnormalities that are associated with psychopathology. It appears that the major forms of hereditary transmission are represented among the disorders with significant psychologic components. The mode of transmission determines the risk, not only for the offspring of affected individuals, should there be such offspring, but also for sibs and their progeny, as well as the progeny of collateral relatives. It is important, therefore, that psychiatrists be familiar with basic genetic principles. It is equally important for psychiatrists to be aware that a genetically determined predisposition rarely signifies the inevitable development of a trait. There are instances already where we can intervene to prevent the noxious consequences of a given genetic constitution and, if not prevent them, ameliorate them or cure the ensuing disease. Furthermore, there is every reason to believe that in those disorders where intervention and prevention are not yet feasible, they will become so in the not too distant future. Meanwhile, genetic counselors face very special problems in the case of psychopathology, where, in addition to the problems existing in counseling parents of any child with a genetic defect, there is the danger of initiating self-fulfilling prophecies. Against this danger has to be weighed the effect of the deceit we practice when we withhold information from other human beings, particularly when that information may influence their decisions regarding childbearing, childrearing, and life-style. These and other questions will be discussed in the succeeding chapters.

REFERENCES

1. Watson, J. B.: *Behaviorism*. New York: Peoples Institute Publishing, 1924.

2. Mendel, G.: Experiments in plant hybridization. In Peters, J. A. (ed.): *Classic Papers in Genetics*. New Jersey: Prentice-Hall, 1959, p. 1.

3. Tschermak, E.: Über kunstliche Kreuzung bei Pisum sativum. *Berichte der Deutschen Botanischen Gesellschaft 18:*232, 1900.

4. De Vries, H.: Sur la loi de disjonction des hybrides. *Comptes rendus de l'Academie des Sciences (Paris) 130:*845, 1900.

5. Correns, C.: G. Mendel's Regel über das Verhalten der Nachkommenschaft der Rassenbastarde. *Berichte der Deutschen Botanischen Gesellschaft 18:*158, 1900.

6. Waters, C. O.: Letter on chorea. In Dunglison, R. (ed.): *The Practice of Medicine*. Philadelphia: Lea and Blanchard, 1848, vol. 2, p. 216.

7. Huntington, G.: On chorea. *Med. Surg. Rept. 26:*317, 1872.

8. Myrianthopoulos, N. C.: Huntington's chorea. *J. Med. Genet. 3:*298, 1966.

9. Van Putten, T. and Menkes, J. H.: Huntington's disease masquerading as chronic schizophrenia. *Dis. Nerv. Syst. 34:54,* 1973.

10. Heathfield, K. W. G. and Mackenzie, I. C. K.: Huntington's chorea in Bedfordshire, England. *Guys Hosp. Rep. 120:*295, 1971.

11. Falek, A. and Glanville, E. V.: Investigation of genetic carriers. In Kallmann, F. J. (ed.): *Expanding Goals of Genetics in Psychiatry*. New York: Grune and Stratton, 1962, p. 136.

12. Klawans, H. L., Paulson, G. W., Ringel, S. P., and Barbeau, A.: Use of L-dopa in the detection of presymptomatic Huntington's chorea. *N. Engl. J. Med. 286:*1332, 1972.

13. Wilson, S. A. K.: Progressive lenticular degeneration: A familial nervous disease associated with cirrhosis of the liver. *Brain 34:*295, 1912.

14. Jarvik, L. F.: Genetic aspects of aging. In Rossman, I. (ed.): *Clinical Geriatrics*. Philadelphia: J. B. Lippincott, 1971, p. 85.

15. Lesch, M. and Nyhan, W. L.: A familial disorder of uric acid metabolism and central nervous system function. *Am. J. Med. 36:*561, 1964.

16. Seegmiller, J. E., Rosenbloom, F. M., and Kelley, W. N.: Enzyme defect associated with a sex-linked human neurological disorder and excessive purine synthesis. *Science 155:*1682, 1967.

17. Kallmann, F. J.: *Heredity in Health and Mental Disorder*. New York: Norton, 1953.

18. Winokur, G. and Tanna, V. L.: Possible role of X-linked dominant factor in manic-depressive disease. *Dis. Nerv. Syst. 30:*89, 1969.

19. Mendlewicz, J., Fleiss, J. L., and Fieve, R. R.: Evidence for X-linkage in the transmission of manic-depressive illness. *JAMA 222:*1624, 1972.

20. Perris, C.: Genetic transmission of depressive psychoses. *Acta. Psychiatr. Scand. (suppl.) 203:*45, 1968.

21. Mendlewicz, J., Fieve, R. R., and Stallone, F.: Relationship between the effectiveness of lithium therapy and family history. *Am. J. Psychiatry 130:*1011, 1973.

22. Stallone, F., Shelley, E., Mendlewicz, J., and Fieve, R. R.: The use of lithium in affective disorders, III: a double-blind study of prophylaxis in bipolar illness. *Am. J. Psychiatry 130:*1006, 1973.

23. Baastrup, P. C. and Schou, M.: Lithium as a prophylactic agent: Its effects against recurrent depressions and manic-depressive psychosis. *Arch. Gen. Psychiatry 16:*162, 1967.

24. Fieve, R. R.: Lithium in psychiatry. *Int. J. Psychiatry 9:*375, 1970–71.

25. Erlenmeyer-Kimling, L., Rainer, J. D., and Kallmann, F. J.: Current reproductive trends in schizophrenia. In Hoch, P. H. and Zubin, J. (eds.): *Psychopathology of Schizophrenia.* New York: Grune & Stratton, 1966, p. 252.

26. Böök, J.: A genetic and neuropsychiatric investigation of a North Swedish population. *Acta Genet. Statist. Med. 4:*1, 1953.

27. Böök, J.: Schizophrenia as a gene mutation. *Acta Genet. Statist. Med. 4:*133, 1953.

28. Elston, R. C. and Campbell, M. A.: Schizophrenia: Evidence for the major gene hypothesis. *Behav. Genet. 1:*3, 1970.

29. Garrone, G.: Étude statistique et génétique de la schizophrenie à Genève de 1901 à 1950. *J. Genet. Hum. 7:*189, 1962.

30. Heston, L. L.: The genetics of schizophrenic and schizoid disease. *Science 167:*249, 1970.

31. Kallmann, F. J.: *The Genetics of Schizophrenia.* New York: Augustin, 1938.

32. Kallmann, F. J.: The genetic theory of schizophrenia: An analysis of 691 schizophrenic twin index families. *Am. J. Psychiatry 103:*309, 1946.

33. Slater, E.: A monogenic theory of schizophrenia. *Acta Genet. Statist. Med. 8:*50, 1958.

34. Burch, P. R. J.: Schizophrenia: Some new aetiological considerations. *Br. J. Psychiatry 110:*818, 1964.

35. Karlsson, J. L.: A hereditary mechanism for schizophrenia based on two separate genes, one dominant, the other recessive. *Hereditas 51:*74, 1964.

36. Rudin, E.: *Zur Vererbung und Neuentstehung der Dementia Praecox.* Berlin: Springer, 1916.

37. Edwards, J. H.: The simulation of Mendelism. *Acta Genet. Statist. Med. 10:*63, 1960.

38. Gottesman, I. I. and Shields, J.: A polygenic theory of schizophrenia. *Proc. Natl. Acad. Sci. USA 58:*199, 1967.

39. Kety, S. S., Rosenthal, D., Wender, P. H., and Schulsinger, F.: The types and prevalence of mental illness in the biological and adoptive families of adopted schizophrenics. In Rosenthal, D. and Kety, S. S. (eds.): *The Transmission of Schizophrenia.* Oxford: Pergamon, 1968, p. 345.

40. Kringlen, E.: An epidemiological-clinical twin study on schizophrenia. Ibid., p. 49.

41. Erlenmeyer-Kimling, L. and Paradowski, W.: Selection and schizophrenia. *Am. Naturalist 100:*651, 1966.

42. Hamburg, D. A.: Genetics of adreno-cortical hormone metabolism in relation to psychological stress. In Hirsch, J. (ed.): *Behavior-Genetic Analysis.* New York: McGraw-Hill, 1967, p. 154.

43. Kidd, K. K. and Cavalli-Sforza, L. L.: An analysis of the genetics of schizophrenia. *Soc. Biol. 20:*254, 1973.

44. Scarr-Salapatek, S.: The heritable nature of group differences: Methodological considerations. Third Annual Behavior Genetics Association Meeting. Chapel Hill, NC, 1973.

45. Heston, L. L.: Psychiatric disorders in foster home reared children of schizophrenic mothers. *Br. J. Psychiatry 112:*819, 1966.

46. Kallmann, F. J.: The genetic aspects of mental disorders in the aging. *J. Hered. 43:*89, 1952.

47. Roth, M.: Interaction of genetic and environmental factors in the causation of schizophrenia. In Richter, D. (ed.): *Schizophrenia: Somatic Aspects.* New York: Macmillan, 1957.

48. Hayward, M. D. and Cameron, A. H.: Triple mosaicism of the sex chromosomes in Turner's syndrome and Hirschsprung's disease. *Lancet 2:*623, 1961.

49. Lejeune, J., Lafourcade, J., Berger, R., et al.: Trois cas de deletion partielle due bras court d'un chromosome 5. *C. R. Acad. Sci. (D) (Paris) 257:*3098, 1963.

50. Edwards, J. H., Harnden, D. G., Cameron, A. H., et al: A new trisomic syndrome. *Lancet 1:*787, 1960.

51. Magenis, R. E., Hecht, F., and Milham, S., Jr.: Trisomy 13 (D1) syndrome: Studies on parental age, sex ratio and survival. *J. Pediatr. 73:*222, 1968.

52. Patau, K., Smith, D. W., Therman, E., et al: Multiple congenital anomaly caused by an extra autosome. *Lancet 1:*790, 1960.

53. Weber, W. W., Mamunes, P., Day, R., and Miller, P.: Trisomy, 17–18 (E): Studies in long-term survival with report of two autopsied cases. *Pediatrics 34:*533, 1964.

54. Turner, H. H.: A syndrome of infantilism, congenital webbed neck and cubitus valgus. *Endocrinology 23:*566, 1938.

55. Ullrich, O.: Über typische Kombinationsbilder multipler Abartungen. *Z. Kinderheilk 49:*27, 1930.

56. Levine, H.: *Clinical Cytogenetics.* Boston: Little, Brown, 1971.

57. Hauser, G. A.: Gonadal dysgenesis. In Overzier, C. (ed.): *Intersexuality.* New York: Academic Press, 1963, p. 298.

58. Hamerton, J. L.: *Human Cytogenetics II. Clinical Cytogenetics.* New York: Academic Press, 1971, pp. 1, 98, 106.

59. Bartalos, M. and Baramki, T. A.: *Medical Cytogenetics.* Baltimore: Williams and Wilkins, 1967.

60. Alexander, D. and Money, J.: Turner's syndrome and Gerstmann's syndrome: Neuropsychologic comparisons. *Neuropsychologia 4:*265, 1966.

61. Garron, D. C.: Sex-linked recessive inheritance of spatial and numerical abilities, and Turner's syndrome. *Psychol. Rev. 77:*147, 1970.

62. Stafford, R. E.: Sex differences in spatial visualization as evidence of sex-linked inheritance. *Percept. Mot. Skills 13:*428, 1961.

63. McClearn, G. E.: Psychological research and behavioral phenotypes. In Spuhler, J. N. (ed.): *Genetic Diversity and Human Behavior.* Chicago: Aldine Press, 1967, p. 33.

64. Hartlage, I. C.: Sex-linked inheritance of spatial ability. *Percept. Mot. Skills 31:*610, 1970.

65. Bock, R. C. and Kolakowski, D.: Further evidence of sex-linked major-gene influence on human spatial visualizing ability. *Am. J. Hum. Genet. 25:*1, 1973.

66. Money, J. and Mittenthal, S.: Lack of personality pathology in Turner's syndrome: Relation to cytogenetics, hormones and physique. *Behav. Genet. 1:*43, 1970.

67. Maclean, N., Harnden, D. G., Court Brown, W. M., et al: Sex chromosome abnormalities in newborn babies. *Lancet 1:*286, 1964.

68. Jacobs, P. A., Harnden, D. G., Buckton, K. E., et al: Cytogenetic studies in primary amenorrhea. *Lancet 1:*1183, 1961.

69. Klinefelter, H. F., Reifenstein, E. C., and Albright, F.: Syndrome characterized by gynecomastia, aspermatogenesis without A-Leydigism, and increased excretion of follicle-stimulating hormone. *J. Clin. Endocrinol. 2:*615, 1942.

70. Tumba, A.: Le phenotype XXXXY étude analytique et synthetique à propos de trois cas personnels et de 67 autres cas de la litterature. *J. Genet. Hum. 20:*9, 1972.

71. Jacobs, P. A., Brunton, M., Melville, M. M., et al: Aggressive behavior, mental subnormality and the XYY male. *Nature 208:*1351, 1965.

72. Hamilton, J. B., Walter, R. O., Daniel, R. M., and Mestler, G. E.: Competition for mating between ordinary and supermale Japanese medaka fish. *Anim. Behav. 17:*168, 1969.

73. Jarvik, L. F., Klodin, V., and Matsuyama, S. S.: Human aggression and the extra Y chromosome—fact or fantasy? *Am. J. Psychol. 28:*674, 1973.

74. Fujita, H., Yoshida, Y., Tanigawa, Y., et al: A survey of sex chromosome anomalies among normal children and mental defectives. *Jap. J. Hum. Genet. 16:*198, 1972.

75. Penrose, L. S.: *The Biology of Mental Defect.* New York: Grune & Stratton, 1963.

76. Smith, P. G. and Jacobs, P. A.: Incidence studies of constitutional chromosome abnormalities in the post-natal population. In Jacobs, P. A., Price, W. H., and Law, P. (eds.): *Human Population Cytogenetics.* Baltimore: Williams and Wilkins, 1970, p. 159.

77. Turner, J. H. and Wald, N.: Chromosome patterns in a general neonatal population. Ibid., p. 153.

78. Money, J.: XYY, the law and forensic moral philosophy. *J. Nerv. Ment. Dis. 149:*309, 1969.

79. Hook, E. B.: Behavioral implications of the human XYY genotype. *Science 179:*139, 1973.

80. Baughman, F. A. and Mann, J. D.: Ascertainment of seven YY males in a private neurology practice. *JAMA 222:*446, 1972.

81. Sumner, A. T., Robinson, J. A., and Evans, H. J.: Distinguishing between X, Y and YY-bearing human spermatozoa by fluorescence and DNA content. *Nature (New Biol.) 229:*231, 1971.

82. Owen, D. R.: The XYY male: A review. *Psychol. Bull. 78:*209, 1972.

83. Court Brown, W. M.: Males with an XYY sex chromosome complement. *J. Med. Genet. 5:*341, 1968.

2

GENETICS, CONSTITUTION, AND GENDER DISORDER

Robert J. Stoller, M.D.

This chapter is a summary of conjectures, hypotheses, data, and rules that presently underlie the management of patients with disorders in the development and maintenance of gender identity—that is, masculinity or femininity. Although massive brain or body malformations are not involved, as they are in other psychiatric genetic disorders, such cases do involve large, relatively abstract scientific issues, especially those concerned with identity development and the mind-brain problem, which influence concrete treatment decisions. For instance, we know that animals range on a scale according to the degree to which their behavior is flexible or drive-bound. The most flexible animal is man, about whom we want to know: When and how much does sex (biology) command gender behavior (psychology), rather than subserve it? Because this issue is crucial in the study of behavior, it shall be emphasized, even though it does not have as much practical use as do the humbler problems of patient management. While correct answers to this question will not in themselves determine treatment plans, they will contribute to the formulation of such plans.

Two basic rules, briefly outlined rather than intensely elaborated or defended, apply to the development of masculinity and femininity in humans:

(1) *The resting state of all mammalian (including human) tissue is female.* If a fetus is to become anatomically and physiologically male, androgen must be added not only to the body but to the brain, which will not "organize" for subsequent masculine behavior without it. These normal processes are under chromosomal (genetic) control. For instance, with only one sex chromosome (XO), no maleness develops, nor does masculine behavior. The second X chromosome is necessary for complete anatomic and physiologic femaleness (but not for femininity). If a Y chromosome is added, anatomic and physiologic maleness is induced via the prenatal production of androgens; thus, the fetus becomes a biologically normal male (XY). In the case of excess chromosomes, abnormalities of sex occur in the predictable direction. For instance, the XXX individual does not develop male characteristics, and the XXY individual is phenotypically male but with some tissue feminization (e.g., gynecomastia).

It seems that early in fetal life the Y chromosome, in an as yet unknown manner, triggers the production of an androgen, which then stimulates adjacent cells to partake in the process, themselves producing increasing amounts of androgens. As these begin circulating, susceptible fetal tissues respond with masculinization, which in time results in the anatomic state described by the term "male." In the absence of this androgenization, a state of femaleness perists.

Among the tissues susceptible to androgens is the brain. When circulating fetal androgens are present, the mammalian brain is modified (in some way not yet known), so that postnatally the animal is capable of masculine behavior. In their absence, feminine behavior, not masculine, occurs. (The terms "feminine" and "masculine," with their connotations of gender behavior, role, and identity, are applicable only to species with complex, flexible, choice-ridden behavior.) But female hormones need not be present for feminine behavior to result. Thus the conclusion that the resting state of mammalian tissue is on the female side, and that maleness and masculinity require the addition of androgens in fetal life and paranatally. (A fine, detailed review of this process is found in Money and Ehrhardt.[1])

(2) *The typical organization of the central nervous system in a male or female direction need not inevitably set future behavior.* Once these

anlagen are laid down, the behaviors they underlie must be supported by the postnatal effects of rearing. And the normal biologic contributions to masculinity and femininity can be disrupted, in fact thoroughly overturned, by environmental events alone.[1,2]

Etiologies of Gender Disorders

Let us review first those (probably) genetic conditions that tend to confirm that the following rule, now well established in animals, applies also to humans: Maleness and masculinity result when androgens are added to tissue, and femaleness and femininity occur in the absence of androgenization. (This work, especially the result of Money's research, is elaborated in Money and Ehrhardt.[1])

Turner syndrome. Lacking a second sex chromosome, children with Turner syndrome are born with normal-appearing female external genitalia and so are assigned to the female sex. Normal ovarian development requires the second X chromosome, and since it is missing in these girls, they will not develop female secondary sex characteristics, which are dependent on ovarian hormones. There are no significant levels of androgens to cause masculinization. These girls are feminine and heterosexual.

Androgen insensitivity syndrome. Although individuals with androgen insensitivity syndrome are XY, their tissues are unable to respond to the testosterone produced by their testes in fetal life, probably because of a genetic defect. As a result, their external genitalia are normally female-appearing (testes, though present, are cryptorchid), and so they are assigned to the female sex and raised as girls. Throughout life their tissue is resistant to testosterone, although the latter is produced in normal amounts. Like normal males, these people secrete a small quantity of estrogen, which, when androgens are missing, is sufficient to produce normal-appearing secondary sex characteristics. These girls are feminine and heterosexual.

Constitutional male hypogonadism. Constitutional male hypogonadism exists in a variety of forms, the commonest of which is Klinefelter syndrome (XXY), usually present with phenotypic maleness but with dysfunctional testes. The exact nature of this gonadal disorder is unknown, but it is frequently found with somatic feminization such as gynecomastia. There is suggestive evidence that a larger than chance number of such males suffer varying degrees of gender reversal from childhood on, apparently as a result of a disturbed fetal CNS androgenization.

Adrenogenital syndrome. In the adrenogenital syndrome, the fetal adrenals produce excessive androgens in subjects who are otherwise apparently normal females. The result is a masculinization of the external genitalia and, it seems, mild masculinizing of the CNS. The evidence for the latter is that a group of these girls was found to be more tomboyish than a control series, although the subjects were not markedly masculine in their behavior or homosexual in object choice.

Although I am presenting no new information, I wish, nonetheless, to comment further about these groups of patients, who raise the question of whether a "biologic force" * may contribute to marked aberrations in masculinity and femininity. Their lives tell us that there are humans in whom powerful biologic factors, stronger than those present in normals, override the psychosocial forces that usually dominate in the development of human gender identity. (But there is no evidence a skeptic should accept proving that gender behavior in most humans is, as in these rare cases, almost solely the result of biologic forces.)

GENDER CLASSIFICATION AND GENDER IDENTITY

In 1964, I began describing people born and assigned to the sex unquestionably indicated by their anatomic appearance, who nonetheless insisted from early childhood that they should be members of the opposite sex.[3] Then, later in life, proper tests established that, except for

* The term "biologic force" is intentionally vague. By no means is this term used to imply that there is some supraordinate quality, of mysterious origin and function, that guides behavior. In the past, areas unknown to science have been prematurely considered understood when encompassed by fancy "biologic" terms, such as "life force" or "libido" or "orgone." All I mean by "biologic force" is that biologic variables not yet clearly delineated are apparently energizing a piece of behavior.

genital appearance, these people had a chromosomal and/or hormonal defect congruent with the sex to which they wanted to belong and opposite from that to which they had been forced to conform.

The first case was that of a child raised from birth as a girl because of normal-appearing female genitalia, but with an otherwise normal male biology (unknown to anyone). In the face of rearing that should have produced a feminine enough girl, the child, with no trace of femininity—to the despair of her parents—forever insisted on being treated like a boy. At age 14, "she" was finally discovered to be in fact a male, biologically normal in all ways except for female-appearing exterior genitalia (the testes were cryptorchid). Since then he has been a masculine young man, and he was recently married.

In writing on this subject in 1968, Baker and I reported on six more patients with gender reversal, males who suffered from conditions that were known to be present since birth but were not detected until maturity.[4] Each individual was born, so far as could be determined when we saw him as an adult, an apparently anatomically normal male, and only years later was it discovered that his testes were inadequate; this was not known in childhood. While the evidence regarding these patients' rearing was fragmentary, in none was there a hint of the parental effects that produce comparable gender reversal in biologic normals—transsexuals. In addition, we noted an unexpected number of reports in the literature describing anatomic males with hypogonadism as having cross-gender behavior from childhood (ranging from a strongly transsexual picture to less feminine activities, such as homosexual preferences or infrequent cross-dressing without excitement [5]; Money and Pollitt had seen this, in regard to hypogonadal males, before we did [6]). Such cases of gender disorder in hypogonadal males continue to be reported, and more so than in other chromosomal, hormonal, or anatomic sex disorders. So far, the outstanding characteristic of such cases is the absence of reported family influences comparable to those to be noted in transsexuals. (The published reports still do not make clear, however, how many of these patients had normal-appearing genitalia at birth; if the genitalia were abnormal, then cross-gender behavior is not unusual and can be accounted for by parental uncertainty.)

Obviously, one cannot perform experiments on humans as one can on animals, and so we must be content with such incomplete evidence as that cited above. At present one can say, nonetheless, that no genetic or other innate disorders have appeared in which gender identity develops contrary to the rules that apply to other mammals. But are these genetic-hormonal-CNS effects all that is necessary for the development of gender identity?

That question brings us to the second rule offered earlier: After the genetically induced anlagen have been laid down, they can almost always be disrupted and even overturned by environmental events alone.

Let us look at two relatively "clean" situations, or "natural experiments," that advance the argument. In the first, the individuals are otherwise intact XY males who have hermaphroditic external genitalia; their appearances range from normally female to more or less male. When unequivocally assigned to a sex—to either sex, for some are considered males and others females—and when the family agrees, the child develops an appropriate gender identity. If, however, the assignment is equivocal, and the parents believe that the child is neither male nor female or that it is both, then a hermaphroditic identity results, in which the child's gender identity, like the assignment, is equivocal. For instance, if the external genitalia look unremarkably female, the child is assigned as a female and grows up having no question of her femaleness and with a gender identity as feminine as can be expected in that family's ambiance. If a child with an identical biologic state happens to be assigned as a male, however, then a clear-cut sense of maleness and masculinity develops. If the parents are told—or for any reason come to believe—that their child is a hermaphrodite, then the child will grow up with the feeling of being different from everyone else in the world and of being some sort of hermaphrodite. In any event, sex assignment, not biology, sets the direction of these subjects' identity development.[7,8]

TRANSSEXUALISM

The second "natural experiment" used herein to show the power of rearing over the biologic condition is the transsexual, who, I shall declare, has, in the absence of any known innate disorder, undergone a complete reversal of gender identity. Male transsexuals become feminine "women" and female transsexuals masculine "men" because of aberrant family dynamics. In these cases, too, parental influences, not biology, cause the reversal.[2] Transsexualism is by definition the most marked reversal of gender identity possible, starting in the first year or two of life, as soon as any behavior that could be called masculine or feminine appears. It thus provides a worthy test of the above-mentioned second rule that postnatal environmental effects can overturn a biologic fundament. And so I shall examine it at greater length, though for brevity let us consider only male transsexualism. (For a hypothesis concerning the etiology of female transsexualism, see Stoller [9].)

Transsexualism may serve our present purposes well, for it exemplifies problems in theory and practice that arise when one tries to determine to what extent an aspect of personality results from biologic and/or environmental forces. We must find a logic to support the needed methodology. And so I shall bear down quite heavily on the present state of research on the etiology of transsexualism, letting this discussion exemplify the issues that trouble us whenever clear-cut biologic defects are not present.

First, a few matters of definition. The term "transsexual" nowadays is used loosely and incorrectly to refer simply to a male or female who requests that his or her body be "transformed" into that of the opposite sex. Within this broadly defined category, however, are diagnostically different types—different in manifest clinical appearance, underlying dynamics, and etiology. Discussions of etiology will be hopelessly blurred if one seeks to find consistent factors in such a mixed group. If, instead, one separates out a clinically homogeneous group—those who are the most feminine of all males, who have been feminine since any gender behavior appeared, and who have had no episodes of masculinity—one will find the following etiologic factors.

The transsexual male's mother is a woman who from childhood has felt hopeless and empty, primarily because her own mother considered her—in the child's femaleness and femininity—worthless. Her father, on the other hand, was at one time close to her, but in that intimacy he encouraged her to adopt his own masculine ways. When, sometime before the child's puberty, he abandons her (by divorce or separation, death, or loss of interest in her), she decides she wishes to be a male.

Then for several years she acts as if she were a female transsexual, dressing only in boys' clothes, playing boys' games only with boys, and believing she will grow a penis in time. Unlike female transsexuals, however, she is discouraged by the changes of puberty—breasts and menstruation—which smash her hopes, and she goes back to a more feminine disguise. In time, she marries.

The man she chooses is a distant, passive, though usually not effeminate man, chosen for the role he will evermore take in the family: to be available for his wife's scorn, but not to be a physical presence whose masculinity would provide strength in the family or a model for his son's developing masculinity.

When the son who is to be transsexual is born—and only a male infant whom this mother considers beautiful and graceful sets off the process *—she is very happy. Her sense of hopelessness and worthlessness ends when she has accomplished this wonderful act of growing from her own body a lovely phallus. Then begins the process that creates the extreme femininity: she holds this infant against her body, day and night, for months and then for years, in a blissful intimacy in which his beauty and his joy in being loved in this way match her yearning for such a loving infant. The two form a symbiosis in which everyone else is essentially excluded, and this mother will do all that is necessary to maintain the intimacy against disruptive forces that might separate the two. As feminine behavior appears, she encourages it and discourages behavior she considers masculine, i.e., forceful, penetrating, dirty, aggressive. (The data behind this sketch are reported in several places; one reference will serve now [2].)

So far, these family dynamics have been found in almost all situations in which the boy was extremely feminine and had had no episodes of masculinity (15 families with the dynamics, 2 without). In each case in which the foregoing clinical picture was not present, the family dynamics were also not present (e.g., in the families of fetishistic cross-dressers, homosexuals, or psychotics with gender disorders). There is

* This seems the one innate factor in the constellation: the infant's beauty, grace, and cuddliness. A more aggressive infant would trouble such a mother and would not spark her potential for establishing an excessively close symbiosis. We should also remember that many pretty, cuddly boys are born who do not become transsexuals.

reason, therefore, to be cautiously confident that the findings will hold up. But this optimism must be fully tempered by the fact that other workers have not affirmed these data. Almost always when etiology is mentioned in discussions of male transsexualism, it is said that the etiology is unknown but that biologic factors must be powerfully present. Now that they have become familiar with the animal research on hormonal reversal of gender behavior, most workers assume that such factors are essential in causing transsexualism. The present discussion presents no proof that their idea is wrong (though that is my belief), but is rather an attempt to question its underlying logic in the hope that more useful and pertinent questions can be asked.

First, let us remember that no one claims that the theory of hormonal reversal is proved, even those convinced of its validity. Their conviction seems to come from four sources: (1) One is often impressed by how profoundly the transsexual's gender identity is fixed, how strong the urge is to change sex, how early in life it started, and how successfully this person manages the role of the opposite sex. (2) Animal work has demonstrated in each species so far tested that adult behavior can be modified in the direction of the opposite sex by manipulation of paranatal hormonal variables. (3) Certain hormonal disorders in humans result in behavior that moves in the directions predicted from the animal work. (4) Biologic theories of behavior have appeal because they seem more scientific, more open to experimentation.

To carry my argument along, it may help us next to recall that three kinds of factors contribute to gender identity: biologic, psychologic, and "biopsychic" (e.g., imprinting, conditioning). In most humans, the development of sex and gender identity are congruent; maleness and masculinity go well together, as do femaleness and femininity. But if the strength of any one of these three factors increases unduly, either prenatally or in infancy-childhood, gender aberration may result. We know of data showing that the norm is breached in animals and/or man in certain "experiments": in the biologic sector, by chromosomal disorders, pre-, peri-, and postnatal endocrine disorders, and surgical and hormonal invasions by physicians; in the psychologic sector, by powerful, aberrant parental, sib, and peer influences; and in a "biopsychic" sector which bridges the other two, by imprinting, classic or visceral conditioning, and explosively traumatic experiences. It is a researcher's task, in studying such disorders, to find out which sector was unduly strengthened (or weakened). In the case of transsexualism, I found these data in the powerful family dynamics. Nevertheless, we know that in animal experiments and genetic-hormonal disorders, a change in the strength of one of the other sectors also causes gender reversal. The experimenter is tempted, therefore, to apply his solid data from *X* (e.g., behavior reversal in male rats given perinatal estrogens) as an explanation for *Y* (e.g., extreme femininity in boys). Such creative postulation *can* lead to the discovery that transsexualism is a paranatal hormonal disorder, but not without confirmatory data.

I am approaching an idea that is simple enough and yet, because it has two parts, is apparently still too complex for acceptance: (1) Transsexualism is the result of family influences, regardless of the fact that gender identity reversal can also be produced experimentally in animals by purely biologic forces. (2) In rare cases, for reasons still not identified, gender identity reversal can *also* be produced in humans by biologic forces, such as congenital hypogonadism in males, regardless of parental effects. In other words, there are two very different causes for reversal. (It is unclear to me whether the clinical pictures in these two types of gender reversals are alike; I have studied only seven patients in depth whose gender disorder was associated with a sex disorder. Two of these resembled transsexuals. The rest, while aberrant and requesting sex change, had features not found in transsexuals, e.g., foot fetishism.)

The animal work gives the strongest impetus to the belief that transsexualism is biologically induced. May there not be some as yet unmeasured hormonal influence at the root of transsexualism? * If so, the

* Money has a biologic explanation that encompasses all kinds and degrees of gender reversal (pp. 251–252 [10]): "The etiology of transexualism though much speculated about, remains essentially unknown. . . . Whereas there are no known laboratory tests available today that show transexuals, as a group, to be consistently physically different chromosomally, hormonally, or morphologically, from a randomly sampled group, there also are no mental formulae or tests that show them to be consistently different in terms of psychodynamic history. [Since this is a refutation of my data and thesis, I miss learning more specifically why Money says it.] Such lack of discrimination may be a function of the crudeness of today's tests. It is possible, for example, that tomorrow's tests will yield

hormone induction theory will have to explain the following obscurities (for which the family dynamics theory has answers in data already collected): Why does transsexualism almost invariably occur in only one child in a family? If a maternal defect releases the hormones that afflict the fetus, the defect ought to occur on occasion in more than one pregnancy; or, if it is a genetic defect passed on to the infant from mother or father, then why is this genetic defect only found in one child of the family? Why is it not seen in uncles, aunts, and cousins, as well as in sibs? In fact, when a family was finally found in which there were two transsexual sibs, the typical family dynamics had existed in the infancy of both children.[11] And a study of two pairs of identical twins, in which one of each pair had suffered gender reversal and the other had not, revealed that only the twins reared by the parents in the manner predicted by my "interpersonal" theory were transsexual.[12]

It is true that animal experiments that tamper with fetal brain physiology can produce adult gender reversal, but the absence of reports of

---

measures of the residual influence of fetal sex hormones on the fetal brain. It has already been established . . . in animal experimentation that sex hormones do in fetal life influence the brain, especially those hypothalamic centers that neurohumorally regulate the cyclic or noncyclic function of the pituitary gland, and also those adjacent centers that coregulate sexual behavior patterns. Not only do sex hormones produced by the fetus itself influence its brain, so also do exogenous hormones injected into the pregnant mother. Moreover, exogenous substances which are hormone antagonists will also influence the fetus if injected into the pregnant mother. For instance, if a pregnant animal is injected with androgen in order to masculinize the fetal females, but is also simultaneously treated with barbiturates, then the injected androgen will have no effect. This experimental finding raises a very cogent question as to the possible, but as yet unknown influence, on the neural sexual centers of the fetal brain, of barbiturates and other medications taken by a woman during pregnancy. It is possible that a pregnancy drug effect could handicap a baby in the early postnatal period of psychosexual differentiation, making him or her easily vulnerable to an error of gender identity. A similar handicap could, in the rare instance of Klinefelter (47, XXY) syndrome, be genetically induced, for it has been found that males with this rare syndrome are also transexual more than would be expected by chance alone" (Money and Pollitt 1964 [6]).

Money suggests that hitherto undiscovered psychologic factors may also contribute, once the brain has been thus disordered: he believes that in some cases the biologic contributes more and in others the psychologic, but that in all cases the disordered brain, damaged by pre- or perinatal biologic forces, is a sine qua non. Since he also considers as transsexualism conditions I would diagnose as homosexuality, transvestism, or other perversions, he asks his biologic theory to cover more ground than I would for any etiologic explanation of gender disorders.

completely gender-reversed ("transsexual"), anatomically normal, free-ranging animals can make us doubt whether such experiments tell us much about real-life etiology. Outside of the laboratory one sees marked, consistent cross-sex behavioral changes in an animal only in those rare cases, such as freemartins, where the opposite-sex hormonal influence is so great that it creates hermaphroditic genitalia. While anything is possible, it seems unlikely that data will appear to support a theory of hormone-induced permanent cross-sex behavior in transsexuals, behavior that is unknown in the free-ranging, anatomically intact members of all other species but is present in man. One should extrapolate from the animal studies to man only when nonexperimental animals also are found with profound gender reversal. So the idea that the animal work already done is evidence of the etiology of transsexualism is shaky.

Before being convinced by the hormonal experiments, we should also insert another link in our chain of logic: an attempt to reverse gender behavior in animals by environmental changes, e.g., conditioning or aberrant rearing, while hormones are not the manipulated variables.* So far as I know, such experiments have not been done, but I would think that animal psychologists could invent them. And it will surely be possible, in biologically normal animals, to discover that changes in rearing can produce gender reversal. There are already hints of this in the behaviorally damaged offspring of Harlow's monkeys.[13]

To repeat, in the hope of simplifying the complicated: While there are rare cases in which cross-gender behavior of differing degrees can be caused by sex hormone defects, another whole series of cases are produced by psychologic factors.

In brief, no convincing evidence has yet surfaced for the theory that

* There is a puzzling absence of experimental work attempting to prove or disprove in animals the hypothesis that environmental effects playing upon the biologically intact organism can cause gender reversal. So far as I know, the hundreds of researchers studying the development of gender behavior in animals are trying to produce abnormal gender behavior only by changing the animal's biologic status, not by environmental (learned) effects. I suspect this one-sided experimentation reflects a current prejudice that gender behavior and its disorders are primarily biologic matters. At any rate, it seems not to have occurred to researchers to attempt to produce gender reversal in biologically normal animals by environmental changes alone.

transsexualism is caused by a hormonal defect. Since at present there is no way to test such a theory, we cannot say that it is not valid. However, I believe the burden of proof is now on those who claim this biologic etiology. They have shown that cross-sex behavior can be hormonally induced in animals; they must now show that the parental factors described above are either not present or coincidental. They will have to demonstrate, for instance, that the most feminine males, those without perversions like cross-dressing fetishism (which would indicate that they highly prized their pleasure-producing penises) or any history of masculine behavior, do not typically have an excessively close symbiosis with the mother or an absent father. They must show that the mothers of such males do not have a predilection for attaching themselves excessively to their infants, and that the infants' beauty, which I think sparks the symbiosis, in fact plays no such part. Although those espousing a primarily biologic cause will be able to cite cases in which not every factor I have described is present, they will have to show that the entire constellation of factors I describe is either *not* present or is coincidental.

The evidence for the ''family dynamics'' theory is still fragmentary, but in contrast to most psychodynamic theories, the proposition can be tested. So far there have been no data that would put it in mortal jeopardy.

Management of Innate Gender Disorders

Finally, let us return from the metaphysical issues of brain and behavior to rules for guiding the management of people with clear-cut genetic (anatomic and physiologic) defects.

One's sense of maleness or femaleness becomes fixed by age three or four [2,7] and can scarcely be reversed, even with massive intervention. Therefore, once one's identity is firm, medical decisions should be made so that one's body can be ''sculpted'' to fit that identity; a formed identity cannot be shifted to accommodate even the most imperious physician. The less said to patient or family about genetic or other biologic facts the better, if those data will create in the patient's or parents' minds any thought that the assigned sex was wrong or that the patient does not belong to either of the only two sexes acceptable to mankind. The data—so scientific to the geneticist—are transformed in the family's mind into dread, magic, oracular pronouncements. What to the unempathic scientist is a chromosome is the heavy hand of immutable destiny to the victims: on receiving the genetic information, the patient may feel transformed into a freak, no longer fully human. Those who feel this is an exaggeration have not treated people afflicted with depression, hopelessness, or psychosis as a result of learning such truth. I spent four years, for instance, with a girl who had become psychotic after being told in her teens that her lack of breasts and menses was the result of an absent sex chromosome and that the missing one might have been male.[2]

The overriding rule is that the geneticist should use his information to sustain identity, not change it. In some cases, then, genetic data will be withheld; in others, they will be stressed when used to support unequivocal sex assignment and rearing. This suggests an ethical question: If the treating physician is inept or obtuse, must he nonetheless be given genetic data he is likely to misuse?

No treatment has yet been devised that is sure to cause reversal of gender identity for child or adult. However, before clear-cut gender identity has formed—i.e., before three or four years of age—it can be shaped to conform to anatomy and/or physiology. Certainly, plans for the future sex of the hermaphrodite newborn can be made without concern for gender identity, which does not yet exist; only genetic and anatomic facts need contribute to decision making. Generally, if the external genitalia are badly malformed in a male infant, the wiser choice is to rear the child as a girl, for the surgical and hormonal issues are at present more easily solved in creating a female-appearing body than in trying to create a male.

In all instances, whether those rare ones in which gender identity formation is primarily the effect of innate factors, or the far more frequent cases in which effects of rearing (family dynamics) are primary, the rule for management of these patients should be: *The facts of the psychologic state—that is, of identity—are more important than the genetic or ana-*

tomic facts. Thus, if a clear sense of maleness and masculinity or of femaleness and femininity has already formed, further intervention should aim to strengthen the already present gender identity, not to reverse it. Only when the gender identity is uncertain (hermaphroditic), unformed (in the newborn), or already reversed (as in the transsexual) can the physician suggest sex reassignment without hazard. In other words, except perhaps in the newborn, one's knowledge of the genetic state should play no part in treatment decisions. The geneticist's truth can be destructive.

REFERENCES

1. Money, J. and Ehrhardt, A. A.: *Man and Woman, Boy and Girl*. Baltimore: Johns Hopkins Press, 1972.

2. Stoller, R. J.: *Sex and Gender*. New York: Science House, 1968.

3. Stoller, R. J.: A contribution to the study of gender identity. *Int. J. Psychoanal. 24:*220–226, 1964.

4. Baker, H. J. and Stoller, R. J.: Can a biological force contribute to gender identity? *Am. J. Psychiatry 124:*1653–1658, 1968.

5. Baker, H. J. and Stoller, R. J.: Sexual psychopathology in the hypogonadal male. *Arch. Gen. Psychiatry 18:*631–634, 1968.

6. Money, J. and Pollitt, E.: Cytogenetic and psychosexual ambiguity: Klinefelter's syndrome and transvestism compared. *Arch. Gen. Psychiatry 11:*589–595, 1964.

7. Money, J., Hampson, J. G., and Hampson, J. L.: Hermaphroditism: Recommendations concerning assignment of sex, chage of sex and psychologic management. *Bull. Johns Hopkins Hosp. 97:*284–300, 1955.

8. Stoller, R. J.: The hermaphroditic identity of hermaphrodites. *J. Nerv. Ment. Dis. 139:*453–457, 1964.

9. Stoller, R. J.: Etiological factors in female transsexualism: A first approximation. *Arch. Sex. Behav. 2:*47–64, 1972.

10. Money, J.: Sex reassignment. *Int. J. Psychiatry 9:*249–269, 1970.

11. Stoller, R. J. and Baker, H. J.: Two male transsexuals in one family. *Arch. Sex. Behav. 2:*323–327, 1973.

12. Green, R. and Stoller, R. J.: Two monozygotic (identical) twin pairs discordant for gender identity. *Arch. Sex. Behav. 1:*321–327, 1971.

13. Harlow, H. and Harlow, M.: Learning to love. *Am. Sci. 54:*244–272, 1966.

3

GENETICS OF NEUROSIS AND PERSONALITY DISORDER

John D. Rainer, M.D.

In 1913, Freud wrote: "We divide the causes of neurotic disease into those which the individual himself brings with him into life, and those which life brings to him—that is to say, into constitutional and accidental." [1] In the same vein, Ernest Jones said in 1930:

> Ever since Mendel's work it has been evident that in estimating the relation of heredity to environment in respect to any character, we have first to ascertain the component units in that character; in other words, what actually constitutes an individual gene. . . . By means of psychoanalysis one is enabled to dissect and isolate mental processes to an extent not previously possible, and this must evidently bring us nearer to the primary elements, to the mental genes in terms of which genetic investigations can alone be carried out. . . . the next study to be applied would be one in the field of heredity.[2]

And in 1937, in one of his late writings, Freud expressed the belief that "we have no reason to dispute the existence and importance of primary congenital variations in the ego. A single fact is decisive, namely, that every individual selects only *certain* of the possible defensive mechanisms and invariably employs those which he has selected." [3]

It is clear from these observations that psychodynamic concepts imply the premise that man is selective with respect to important aspects of his life experiences. For that reason, man can be thought of as helping to create his own environment. It follows, therefore, that the complexities of aberrant development and behavior, as delineated with increasing skill by psychodynamic research, will be fully understood only if the genetic dimension is considered as an integral part of the overall frame of reference.

The application of genetics to neurotic illness and personality disorder, and certainly to normal variations in personality, is at a less sophisticated stage than is the case with the major psychoses. Problems of diagnosis and methodology are greatly increased, and there is no convergence of data toward a uniform synthesis. This chapter will illustrate some of the approaches to the coordination of genetics with clinical and behavioral malfunction. The essay is by no means exhaustive; there have been a number of recent reviews [4–6] which need not be duplicated here.

The literature on genetics and nonpsychotic behavior disorders falls into a number of categories. Among these are studies of families or of series of twins; some studies are clinical, some are based on various personality rating scales, and some use psychologic or psychophysiologic tests. Another approach is to consider differences among neonates or infants in baseline attributes and responses to stimuli, and to correlate these with subsequent behavior. Longitudinal or retrospective studies of individual pairs of identical twins may at least offer hypotheses regarding significant factors in the dynamic gene-environment interaction. And finally, the speculations and clinical judgments of psychoanalysts and child psychiatrists who are unafraid to consider the role of inborn factors may furnish the material for more refined biologic investigation. In discussing the comprehensive dynamics of human behavior, Sandor Rado noted: "Psychodynamics must be recognized as the basic component of this structure, because it alone can discover the behavior problems which await their solution from physiology and genetics." [7]

Family studies of neurosis may be typified by the pre–World War II work of Brown, who investigated the relatives of patients with anxiety states, hysteria, and obsessional states, comparing these relatives' risks with those in control populations. He found a significantly increased risk in the patients' first-degree relatives and a tendency for the same type of neurosis in both patient and relatives, although significant overlapping occurred. [8]

In developing a heuristic theory of the neurotic constitution, Slater and Slater [9] included these data, as well as factor analysis of family and personal history, in their study of a large group of soldiers. According to their hypothesis, the constitution might be determined by a large number of genes with additive effect, producing a reduced resistance to some form of environmental stress. Genes with dissimilar effects would be related to different types of stress and produce different neurotic symptoms.[9]

In a later study of twins with diagnoses of neurosis or personality disorder, followed as outpatients at the Maudsley, Shields and Slater found that 29 percent of monozygotic (MZ) co-twins had the same diagnosis as the index case, compared with only 4 percent of dizygotic (DZ) co-twins.[10] Anxiety states showed a higher concordance than other neuroses, while hysteria showed little evidence at all for a specific genetic basis.[11]

Reviewing 15 twin studies of obsessive-compulsive neurosis, Inouye found a total of 27 out of 35 MZ pairs concordant, compared to 0 out of 7 DZ pairs; in 12 studies of hysteria, the evidence for genetic contribution was less, with concordance in only 9 of 42 MZ pairs and 0 out of 43 DZ pairs.[4]

Studies based on psychologic tests have involved general neurotic traits as well as factors within the normal range of personality variation. For example, applying factor analysis to personality characteristics as measured by a wide variety of psychologic techniques, Eysenck observed that many differently formulated theories of neurosis were on the same descriptive level and concerned with the same fundamental dimension of personality.[12] Together with Prell, he was one of the few investigators of psychoneurotic reaction potentials who availed himself of the opportunities afforded by the twin-study method. On the basis of data obtained from 25 one-egg and 25 two-egg pairs, Eysenck and Prell classified the "neurotic personality factor" as a biologic and largely gene-specific entity, estimating the genetic contribution to this "neurotic unit predisposition" at 80 percent.

Another similar approach involving twins is described in a monograph by Claridge, Canter, and Hume in which a sample of 95 pairs was used to test out some of the senior author's hypotheses concerning the psychophysiologic classification of personality.[13] A variety of physiologic measures were analyzed individually, and the data were then subjected to factor analysis. Two principal factors were isolated. One is termed "tonic arousal"; it refers to autonomic reactivity and is made up largely of two measures, heart rate level and sedation threshold. The other factor is termed "arousal modulation"; it is concerned "with the monitoring of sensory input and with such processes as narrowing and broadening of attention," and is mainly associated with EEG parameters, such as alpha index and alpha frequency. It is the latter factor—arousal modulation—that seems to be more influenced by heredity. Both factors form the parameters of Claridge's descriptive model of personality, which has two major dimensions, "neuroticism" and "psychotism." In his rather complex formulation, tonic arousal and arousal modulation are positively associated in neuroticism, the hysteroid being low in both, the obsessoid high in both. In psychotism, on the other hand, Claridge proposes a dissociation, with low tonic arousal and high modulation in the schizoid and the opposite in the cycloid.

Gottesman used the Minnesota Multiphasic Personality Inventory (MMPI) with twins to investigate heritability on the various scales.[14] Social introversion and depression (as well as psychopathic deviation) showed the highest heritability, hypochondriasis and hysteria (together with paranoia and pathologic sexuality) the lowest. Gottesman extrapolated these trait findings and concluded that neuroses with anxiety and schizoid reaction have a high genetic component, given environmental conditions similar to those of his subjects; that those with depressive, phobic, and obsessive-compulsive symptoms have a lower genetic component; and that those with hypochondriac and hysteric elements have little or no genetic component.

If a relationship between personality type and choice of neurosis or character disorder is implicit in investigations using personality scales, it is equally basic to studies of innate differences and personality development in infants and children. Levy was one of the first to call attention to the fact that intrinsic differences among children are equally as important as maternal attitudes in determining future behavior patterns.[15]

Thomas, Chess, and Birch have studied a group of temperamental qualities in young infants which appear to persist into early childhood.[16] Among these are activity level, rhythmicity, approach-withdrawal, adaptability, intensity of reaction, threshold of response to stimulation, quality of mood, distractibility, and attention span and persistence. Escalona has defined eight major areas, or "dimensions," of behavior: activity level, perceptual sensitivity, motility, bodily self-stimulation, spontaneous activity, somatic need states and need gratifications, object-related behavior, and social behavior.[17] And Korner has discussed and elaborated on the importance of studying individual differences in the following neonatal variables: frequency and length of periods of alert inactivity; singular, global, or multiple response to stimuli; response to multiple or competing stimuli; influence of internal state on behavior; distinctness of state; zone reliance; mode reliance; dedifferentiation (regression under tension); and self-consistency.[18]

To be sure, not all constitutional variations are hereditary, since prenatal and perinatal factors are to be considered; subsequently, any inborn or congenital differences will have different valences in different environments. Just as the dietary input of the diabetic or the phenylketonuric will crucially influence his life and development, so the varying opportunities provided by home and school for love and aggression, cooperation and competition, will evoke and facilitate, or fearfully suppress, the individual's capacity for effective action. While one can in theory generalize about the average expectable environment, matched by the average congenital ego equipment, there are many deviations in both factors. One example is provided by the study of deaf children and blind children. The innate sensory defects not only modify the child's perceptive and cognitive abilities but also influence the way his parents accept him and interact with him on the communicative and affective levels. Deaf adolescents may range from relatively normal social adjustment through various degrees of immaturity, impulsivity, and shallowness, while the blind child may develop well or to the extreme of autism. There is evidence that the character traits of children with sensory deprivation depend very much on the specific mother-child setting.

In the psychoanalytic framework, then, it could be said until recently that while dynamic psychiatry had contributed much to the understanding of human development (to "genetic" psychology in the sense that psychologists use the word), the role of hereditary, inborn factors (what biologists would call "genetic") was infrequently associated with psychoanalysis and was even considered antithetic to its principles.

Yet a survey of both psychoanalytic theory and clinical observation will make it clear that the concept of inborn differences plays an essential role in the structure of psychoanalysis as a science, that such differences have been surmised from the analysis of adults and can be observed in studies of the development of children, and that they may have predictive and therapeutic value.

In psychoanalysis, the problems of innate or instinctive versus acquired or learned behavior are of central importance. Freud variously defined instinct as the measure of the demand made upon the mind in consequence of its connection with the body—a borderline concept between the mental and the physical. While this concept has been criticized as nonoperational, it may be useful in separating the source, aim, and object of human drives. Having their source in the temporary states of tissue disequilibrium which occur during metabolic and other life processes, such drives may aim at achieving a homeostatic condition. In pursuing this aim, the individual comes to choose methods which depend on the history of his relations with other persons ("objects") who have been important in helping him. Typical action patterns develop as the drives are directed toward such persons and are adapted to the superstructure of social relationships.

There is no doubt that the study of these processes will require help from chemistry, anatomy, biology, and the social sciences, and there is also no doubt today that genetics plays a role at every level and in every discipline.

Among the individual's inborn psychologic equipment are the potentialities for those functions which Hartmann assigns to the conflict-free ego sphere [19] and which exert their effects outside the region of mental conflicts. These include perception, intention, object response, thinking, language recall, productivity, motor development, maturation, and learning. In addition to these ego functions, a wide variety of studies and opinion express the strong conviction that there are heritable variations in drive strengths and defense mechanisms, as well as in pleasure potential, anxiety proneness (including separation anxiety), and capacities for identification and for postponement of gratification. Freud laid down the concept of *complemental series:* the stronger the hereditary disposition to a form of psychopathology, the less need for life experiences to bring it out. Less mechanical is the notion propounded by Benjamin of the interaction of both the innate and experiential: "Not only can innate differences in drive organization, in ego functions, and in maturational rates determine different responses to objectively identical experiences, but they can also help determine what experiences will be experienced, and how they will be perceived." [20]

Hartmann has suggested using twins in clinical investigation to study the "substitution" of traits, the search for character anlagen and their differentiation into "character *traits.*" [21] Similarly, Korner has discovered that "it is likely that genetic differences in the biological sense may help create genetic differences in the psychoanalytic sense. It is equally plausible that differences in primary endowment may throw a unique cast on the manner in which a child will experience and master each developmental step." [18]

True longitudinal studies, especially of twins, could throw much light on the spiral development from a common origin. Again, in Hartmann's words:

> We are entitled to expect some answer from the twin-study method. The possibilities it opens can be of particular use . . . if we do not limit ourselves to determining whether or not a given trait (in the phenotype) appears in one or both twins—which would of course be sufficient for a biogenetic analysis of certain diseases—but also observe the growth of the personality by means of the interaction between heredity and environment. . . . Hence we must give a special place within this over-all problem to the study of monozygotic twins in their early age.[21]

Retrospective studies are subject to omission and distortion, but intensive investigation of twins and their families may contrast similarities in underlying personality with differences in overt behavior attributable to patterns of family dynamics as well as to physiologic differences. Such studies include the dissimilar sexual orientations of identical twins investigated by Rainer et al.,[22] and the pair of twins concordant for the XYY chromosome anomaly but unlike in degree of control over episodic impulsive behavior, as described by Rainer, Abdullah, and Jarvik.[23]

This description of investigations into the interaction of nature and nurture could be extended to the psychosomatic illnesses, typified by Mirsky's work on peptic ulcer [24] and Spitz's description of the predispositional and precipitating factors in infantile eczema.[25]

In concluding this discussion, it is important to emphasize the role of genetic factors both in the therapy and in the prevention of neurotic illness. As far as treatment is concerned, appreciation of this role will lead not to therapeutic nihilism but to an understanding of the core problems, which may then be dealt with by more pointed and specific psychotherapeutic, biologic, or pharmacologic means. And in the area of prevention, the value of the genetic approach will lie in the better understanding of individual differences as they relate to child rearing and educational practice. In this context, it is usually recommended that parents and teachers tailor their input to the stronger components of the child's intrinsic makeup. Anna Freud, with her usual wisdom, proposes a strikingly different formulation:

> Inherent potentialities of the infant are accelerated in development, or slowed up, according to the mother's involvement with them or the absence of it. Unharmonious progress is balanced out if the parents libidinize lines on which the child lags behind, instead of making the common mistake of giving the highly intelligent children more food for intelligence; talking to the particularly verbal; and giving the bodily active more opportunity for action.[26]

But whether directed at compensating for weak potentials or capitalizing on strong ones, individualizing the environment's contribution to growth and development obviously requires attention to the genotype. Thus, in the larger social sphere, genetics may do its part in reducing neuroticism and increasing the range of effective human functioning.

REFERENCES

1. Freud, S.: The disposition to obsessional neurosis: a contribution to the problem of choice of neurosis. In Strachey, J. (ed.): *Standard Edition of the Complete Psychological Works of Sigmund Freud.* London: Hogarth Press, 1958, vol. 12, pp. 317–326.

2. Jones, E.: Mental heredity. In *Essays in Applied Psychoanalysis.* London: Hogarth Press, 1951, vol. 1, pp. 133–134.

3. Freud, S.: Analysis terminable and unterminable. In Strachey, J. (ed.): *Standard Edition*. London: Hogarth Press, 1964, vol. 23, pp. 216–254.

4. Inouye, E.: Genetic aspects of neurosis: a review. *Int. J. Ment. Health 1:*176–189, 1972.

5. Miner, G. D.: The evidence for genetic components in the neuroses. *Arch. Gen. Psych. 29:*111–118, 1973.

6. Slater, E. and Cowie, V.: *The Genetics of Mental Disorders*. London: Oxford University Press, 1971, chap. 5.

7. Rado, S.: Adaptational psychodynamics, a basic science. In *Psychoanalysis of Behavior*. New York: Grune & Stratton, 1956, vol. 1, pp. 332–346.

8. Brown, F. W.: Heredity in the psychoneuroses. *Proc. R. Soc. Med. 35:*785–790, 1942.

9. Slater, E. and Slater, P.: A heuristic theory of neurosis. *J. Ment. Psychiat. 7:*49–55, 1944.

10. Shields, J. and Slater, E.: Diagnostic similarity in twins with neurosis and personality disorders. In Shields, J. and Gottesman, I. (eds.): *Man, Mind and Heredity*. Baltimore: Johns Hopkins University Press, 1971, chap. 21.

11. Slater, E.: The 35th Maudsley lecture: "Hysteria 311." *J. Ment. Sci. 107:*359–381, 1961.

12. Eysenck, H. J. and Prell, D. B.: The inheritance of neuroticism: An experimental study. *J. Ment. Sci. 97:*441–467, 1951.

13. Claridge, G., Canter, S., and Hume, W. I.: *Personality Differences and Biological Variations: A Study of Twins*. Oxford: Pergamon, 1973.

14. Gottesman, I. I.: Heritability of personality: a demonstration. *Psychol. Monographs 77:*1–21, 1963.

15. Levy, D.: *Maternal Overprotection*. New York: Columbia University Press, 1943.

16. Thomas, A., Chess, S., and Birch, H.: *Behavioral Individuality in Early Childhood*. New York: New York University Press, 1963.

17. Escalona, S. K.: *The Roots of Individuality: Normal Patterns of Development in Infancy*. Chicago: Aldine, 1969.

18. Korner, A. F.: Some hypotheses regarding the significance of individual differences at birth for later development. In *The Psychoanalytic Study of the Child*. New York: International Universities Press, 1964, vol. 19, pp. 58–72.

19. Hartmann, H.: Comments on the psychoanalytic theory of the ego. In *The Psychoanalytic Study of the Child*. New York: International Universities Press, 1950, vol. 5, pp. 74–96.

20. Benjamin, J. D.: The innate and the experiential in development. In Brosin, H. W. (ed.): *Lectures in Experimental Psychiatry*. Pittsburgh: University of Pittsburgh Press, 1961.

21. Hartmann, H.: Psychiatric studies of twins. In *Essays on Ego Psychology*. New York: International University Press, 1964, chap. 20.

22. Rainer, J. D., Mesnikoff, A., Kolb, L. C., and Carr, A.: Homosexuality and heterosexuality in identical twins. *Psychosom. Med. 22:*251–259, 1960.

23. Rainer, J. D., Abdullah, S., and Jarvik, L. F.: XYY karyotype in a pair of monozygotic twins: a 17-year life-history study. *Brit. J. Psychiat. 120:*543–548, 1972.

24. Mirsky, I. A.: Physiologic, psychologic, and social determinants of psychosomatic disorders. *Dis. Nerv. Syst. (monograph, suppl.) 21:*50–56, 1960.

25. Spitz, R.: *The First Year of Life*. New York: International University Press, 1965, pp. 224–242.

26. Freud, A.: *Normality and Pathology in Childhood*. London: Hogarth Press, 1966, p. 233.

4

GENETICS OF AFFECTIVE DISORDERS

Remi J. Cadoret, M.D. and
George Winokur, M.D.

Affective disorder is an extremely common psychiatric condition. Most of the studies of population prevalence have come from Scandinavian countries and have shown rather high rates. Helgason [1] reported a life expectancy for affective disorder in an Icelandic population as 5.2 percent for men and 8.3 percent for women. Sørenson and Strömgren [2] found a point prevalence (incidence of the condition on one day) for depression of 3.9 percent, and Juel-Nielson and co-workers [3] found a 12-month incidence of 0.2 to 0.5 percent in an adult population who contacted an institution for help with depression. However common they may be, affective conditions comprise a heterogeneous group whose similarity lies in the signs and symptoms of the depressive picture. Before dealing with possible genetic etiologies, it is necessary to delimit groups of depressions which are more homogeneous. This has been done as follows: on the basis of preexisting psychiatric conditions other than depression, social and medical conditions, and familial constellations of psychiatric illnesses, affective disorders have been divided by Winokur [4] and by Robins and Guze [5] into those shown in Table 1.

Grief reaction may be considered a model of a reactive depression.

TABLE 1

Classification of Affective Disorder

Type of Affective Disorder	Clinical Manifestations	Preexisting Psychiatric Condition	Preexisting Social Condition	Familial Psychiatric Illness
I. Bereavement or grief reaction	Depression		Recent loss of emotionally meaningful person	
II. Secondary depression	Depression + symptoms of preexisting psychiatric conditions (if still active)	Hysteria, anxiety neurosis, obsessional illness, alcoholism, sociopathy, schizophrenia, organic brain syndrome, sexual deviation	May have recent loss of relationship, eg suicidal alcoholic who has just lost his home, or homosexual who has broken up with sexual partner	Relevant to preexisting psychiatric condition, eg alcoholism in families
III. Primary affective disorders		No preexisting psychiatric condition		
(A) Bipolar illness (manic-depressive disease)	Depression and/or mania	No preexisting psychiatric condition		Mania occurs in family (see Table 2)
(B) Unipolar illness (depressive disease)	Depression only	No preexisting psychiatric condition		No mania in family
(1) Pure depression				Low incidence of alcoholism and sociopathy in family
(2) Depression spectrum disease				High incidence of alcoholism and sociopathy in family

One survey of individuals who had lost a spouse indicated that 35 percent of the bereaved developed a depressive syndrome.[6] No particular family constellation or preexisting psychiatric condition stands out as a characteristic.

Secondary depression by definition occurs during the course of other psychiatric illnesses. Family constellation is related to the preexisting psychiatric condition.

Primary affective disorder is delimited from the above by two features: (1) no preexisting psychiatric condition and (2) no recent bereavement. The clinical picture of primary affective disorder is that of a depression and/or a mania. The criteria for diagnosing mania or depression can be found in Feighner et al.[7] In addition to mania, which distinguishes bipolar from unipolar illness, there are other significant clinical and familial differences between the two conditions. These are shown in Table 2. The familial variables are further evidence of a difference between unipolar and bipolar conditions, and at the same time point to a possible genetic mode of transmission. The higher incidence of two-generation histories in bipolar families suggests a dominant-gene hypothesis, and the different incidences of affected sons and daughters of ill mothers (Table 2, last line) argue for a different kind of transmission compatible with a sex-linked one for bipolar illness but *not* for unipolar illness.

There is direct evidence available that bipolar illness involves an X chromosome or sex-linked transmission. Two groups of investigators [8,9] have reported a significant genetic linkage between bipolar illness and red-green color blindness, which is caused by a gene located on the X chromosome. In two studies,[10,11] one other X-linked character has been associated with bipolar illness: Xg^a blood group. These studies and another [12] which also supports X-linked transmission are shown in Table 3. Four further studies [13–16] which have produced evidence against X linkage are also presented in Table 3. One of the consequences of an X-linked illness is that ill fathers can transmit the condition to their daughters, not their sons. These latter four studies found numerous instances of apparent father-to-son transmission, a fact which by itself is not compatible with an X-linked condition. However, in two of four studies,[14,15] an alternative explanation is possible: a diathesis for

TABLE 2

Clinical and Familial Differences Between Bipolar (Manic-Depressive) Disease and Unipolar (Depressive Disease) Conditions

Variables	Bipolar Affective Disorder	Unipolar Affective Disorder
Clinical:		
(1) Age of onset	Younger (med = 28 yrs.)	Older (med = 36 yrs.)
(2) Bi- or triphasic course of episode (D,M,D), (M,D), or (D,M)	Yes	No
(3) 6 or more episodes of illness	57%	18%
Familial:		
(1) Two-generation history of affective illness (proband-parent; proband-child)	54%	34%
(2) Affective illness in extended family	63%	36%
(3) Bipolar illness in first-degree relatives	10.2–10.8%	.29–.35%
(4) Son vs. daughter risk for affective disorder when mother ill	23% vs. 24%	9% vs. 23%

illness could have come from the maternal side. In the Von Greiff [13] and Green [16] studies, the investigators made special efforts to rule out pathology on the maternal side and still found cases with father-to-son pattern of illness transmission. We are thus left with evidence for two kinds of transmission in bipolar illness: X-linked and possible autosomal. This is analogous to the situation in retinitis pigmentosa, where partial sex linkage accounts for transmission in some families and an autosomal gene accounts for transmission in other families. At the present time we are unable to distinguish the two types of bipolar illness clinically.

Just as the study of families showed different transmission patterns and incidences of illness between unipolar and bipolar types, similar consideration of familial patterns of illness has led to a dichotomy of unipolar affective disorder into early- and late-onset depressions. In the

TABLE 3

Evidence for and against X-Linkage of Bipolar Illness

Study	Method	Findings
Results compatible with X-linked transmission		
Reich, Clayton, and Winokur [8]	Red-green color blindness	In 2 extensive families, highly significant association of carrier state and color blindness.
Winokur and Tanna [10]	Xg^a	In 3 families, 10 children were compatible with sex-linked transmission, one against—a significant difference.
Mendlewicz, Fleiss, and Fieve [11]	Xg^a	Seven families studied. Significant association between blood type and illness.
Fieve et al.[9]	Red-green color blindness	Nine families; close linkage found for proton and deuton and illness.
Taylor and Abrams [12]	Analysis of parent-child pairs—both disordered	Report no ill father-son pairs in study of 55 manic probands.
Results incompatible with X-linked transmission		
Von Greiff, McHugh, and Stokes [13]	Analysis of parent-child pairs, both affectively disordered	In 16 male bipolar patients, 4 had affectively disordered fathers with no evidence of affective disturbance on maternal side.
Perris [14]	Analysis of parent-child pairs, both affectively disordered	Seven father-male proband pairs reported where both were ill.
Dunner, Gershon, and Goodwin [15]	Analysis of parent-child pairs, both affectively disordered	In 23 bipolar male patients, 4 had affectively disordered fathers. Possibility for illness on maternal side.
Green et al [16]	Analysis of parent-child pairs, both affectively disordered	In 35 bipolar families, 4 cases of father-son illness. No evidence of affective illness on maternal side.

early-onset groups, individuals who became depressed prior to age 40 had considerably more alcoholism in the family than did the late-onset depressives (first illness after age 40). Winokur et al.[17] picked out two prototypic groups of probands to represent each kind of depression: early onset represented by females, and late onset by males. Because the early-onset groups have more alcoholism and other conditions, such as sociopathy in male relatives, it has been named "depression spectrum." This term carries the connotation that the increased incidence of alcoholism and personality disorder represents a "depressive equivalent" or a different, sex-influenced phenotype of what may be a similar familial genotype.

Because the late-onset group has so few family members with alcoholism or other psychiatric conditions, it has come to be called the "pure depression" group. This group, typified by late-onset males, shows one other feature differentiating it from the "depressive spectrum" group: *equal* numbers of male and female relatives of late-onset males have affective disorders, in contrast to early-onset females, where *more* female than male relatives suffer from affective disorders.

The differences between the two prototype groups of depression discussed above were first found in a series of 100 depressive inpatients.[17] Since that time, the same differences have been sought in five additional studies. The results of these studies appear in Table 4 and involve over 1,200 probands. For the most part, the differences reported in the original study (line 1) have been replicated in the additional five studies. On the basis of this consistency, it is reasonable to conclude that these groups are different in their family constellations of psychiatric illnesses. This difference has also been extended to clinical variables. Age of onset of depressive disease, even when controlled for the current age of the patient, is related to some symptom differences: early-onset patients have more guilt feelings, fearfulness, and suicide attempts than patients whose onset was after 40 years of age.[22]

The exact mode of transmission of unipolar depressive illness, whether it be depressive spectrum disease or pure depressive type, is not at all clear. From the pattern of familial illness, it is apparent that a sex-linked inheritance is not implicated (see Table 2, last line) and that the

TABLE 4

Familial Psychiatric Illness in Depressive Patients Separated According to Age of Onset and Sex: Consistency Across 6 Studies (1,255 Probands)

Study	No. of probands	More familial AD in EO than LO probands	More familial Alc in EO than LO probands	AD in relatives of LO males; male relatives ≥ female relatives	AD in relatives of EO females; female relatives > male relatives
(1) Winokur et al.[17]	(N = 100)	Yes	Yes	Yes	Yes
(2) Winokur et al.[17]	(N = 345)	Yes	Yes	Yes	Yes
(3) Woodruff, Guze, and Clayton[18]	(N = 139)	Yes	Yes	No	Yes
(4) Winokur[19] *	(N = 242)	Yes	Yes	Yes	Yes
(5) Martin et al.[20]	(N = 204)	Yes	Yes	Yes	Yes
(6) Winokur et al.[21] (Iowa 500)	(N = 225)	Yes	Yes	Yes	No

Note: AD = Affective disorder; Alc = Alcoholism and/or sociopathy

* In this study early onset (EO) is <40, and late onset (LO) >50; in all other studies EO <40, and LO >40.

transmission is different from that found in bipolar illness. We have no linkage studies (as with bipolar illness) that either prove a genetic factor or indicate a specific type of genetic transmission.

The early-onset illness would appear to be sex-influenced in its manifestation: females show affective disorder and males mainly alcoholism and, to a lesser degree, antisocial personality disorder. The total amount of psychiatric illness shown by both sexes is essentially equal [17] and is about 40 percent. This degree of overall risk for relatives is, of course, much higher than the similar risk for relatives of probands with pure depressive disease (12–20 percent).

One survey of published series of late-onset depressives has described an excess of females ill with affective disorder among relatives of late-onset female probands.[20] This finding has at least two interpretations: (1) that there is a third kind of depressive disorder in which more females than males are affected, or (2) that some overlap with early-onset illness could have occurred, with probands with depressive spectrum disease becoming ill later, e.g., in their 5th decade. The latter interpretation implies that the usual dichotomy point of 40 years of age may not be the optimum for separation. Studies are currently under way to determine if this is the best dichotomy in terms of producing homogeneous groups of affective disorders. Because of different ages of onset for male and female depressives,[23,24] it is not unlikely that different ages of dichotomy for the two sexes may eventually give more homogeneous separation.

These studies have shown that primary affective disorder can be broken down into subgroups on the basis of family constellation. Further follow-up of such subgroups should determine the clinical usefulness of the breakdown. In the case of bipolar illness, in families where the mode of transmission is sex-linked, some genetic counseling can be done. With unipolar depressive illness, age of onset of illness gives some idea of increased risk in first-degree relatives for affective disorders and alcoholism. This fact could prove of use in genetic counseling, though there are no long-term follow-up studies of children of early-onset depressives to assess directly the risks for illness.

REFERENCES

1. Helgason, T.: The frequency of depressive states in Iceland as compared with the other Scandinavian countries. *Acta Psychiatr. Scand.* (*suppl. 162*) *37:*81, 1961.

2. Sørenson, A. and Strümgren, E.: Prevalence (the Samso investigation). *Acta Psychiatr. Scand.* (*suppl. 162*) *37:*62, 1961.

3. Juel-Nielson, N., Bille, M., Flygenring, J., and Helgason, T.: Incidence (the Aarhus county investigation). *Acta Psychiatr. Scan.* (*suppl. 162*) *37:*69, 1961.

4. Winokur, G.: The types of affective disorder. *J. Nerv. Ment. Dis. 156:*82–96, 1973.

5. Robins, E. and Guze, S. B.: Classification of affective disorders: the primary-secondary, and endogenous-reactive, and the neurotic-psychotic concepts. In Williams, T. A., et al (eds.): *Recent Advances in the Psychobiology of the Depressive Illnesses: Proceedings of a Workshop Sponsored by NIMH.* Washington, DC: Government Printing Office, 1972.

6. Clayton, P. J., Desmarais, L., and Winokur, G.: A study of normal bereavement. *Am. J. Psychiatry 125:*168–178, 1968.

7. Feighner, J. P., Robins, E., Guze, S. B., et al: Diagnostic criteria for use in psychiatric research. *Arch. Gen. Psychiatry 26:*57–63, 1972.

8. Reich, T., Clayton, P., and Winokur, G.: Family history studies: V. The genetics of mania. *Am. J. Psychiatry 125:*1358–1369, 1969.

9. Fieve, R. R., Mendlewicz, J., Rainer, J. D., and Fleiss, J.: A dominant X-linked factor in manic depressive illness: studies with color blindness. Paper presented at 63rd annual meeting, American Psychopathological Association, New York, 1973.

10. Winokur, G. and Tanna, V. L.: Possible role of X-linked dominant factor in manic-depressive disease. *Dis. Nerv. Syst. 30:*89–93, 1969.

11. Mendlewicz, J., Fleiss, J., and Fieve, R. R.: X-linked dominant transmission in manic-depressive illness (linkage studies with Xg^a blood group). Paper presented at 63rd annual meeting, American Psychopathological Association, New York, 1973.

12. Taylor, M. and Abrams, R.: Manic states: a genetic study of early and late onset affective disorders. *Arch. Gen. Psychiatry 28:*656–658, 1973.

13. Von Greiff, H., McHugh, P. R., and Stokes, P.: The familial history in sixteen males with bipolar manic-depressive disorder. Paper presented at 63rd annual meeting, American Psychopathological Association, New York, 1973.

14. Perris, C.: Genetic transmission of depressive psychoses. *Acta Psychiatr. Scand.* (*suppl. 203*):45–52, 1968.

15. Dunner, D., Gershon, E., and Goodwin, F. K.: Heritable factors in the severity of affective illness. Paper presented at 123rd annual meeting, American Psychiatric Association, San Francisco, 1970.

16. Green, R., Goetze, V., Whybrow, P., and Jackson, R.: X-linked transmission of manic-depressive illness. *JAMA 223:*1289, 1973.

17. Winokur, G., Cadoret, R. J., Dorzab, J., and Baker, M.: Depressive disease: a genetic study. *Arch. Gen. Psychiatr. 24:*135–144, 1971.

18. Woodruff, R., Guze, S., and Clayton, P.: Unipolar and bipolar primary affective disorder. *Br. J. Psychiatry 119:*33–38, 1971.

19. Winokur, G. W.: Diagnostic and genetic aspects of affective illness. *Psychiatric Annals,* Feb. 1973.

20. Martin, S., Cadoret, R. J., Winokur, G., and Ora, E.: Unipolar depression: a family history study. *Biol. Psychiatry 4:*205–213, 1972.

21. Winokur, G., Morrison, J., Clancy, J., and Crowe, R.: The Iowa 500: familial and clinical findings favor two kinds of depressive illness. *Compr. Psychiatry 14:*99–107, 1973.

22. Baker, M., Dorzab, J., Winokur, G., and Cadoret, R.: Depressive disease: classification and clinical characteristics. *Compr. Psychiatry 12:*354–364, 1971.

23. Matusek, P., Halbach, A., and Troeger, V.: *Endogene Depression.* Munich-Berlin: Urban and Schwarzenberg, 1965.

24. Perris, C.: A study of bipolar (manic-depressive) and unipolar recurrent depressive psychosis. *Acta Psychiatr. Scand.* (*suppl. 194*): 1966.

5

GENETICS OF SCHIZOPHRENIA *

Edward H. Liston, M.D. and
Lissy F. Jarvik, M.D., Ph.D.

"Despite various advances in recent years, psychiatric research is still battling on many fronts, in America as elsewhere, for general recognition of genetic concepts. . . . The key position of this battle seems to be held by the disease group of schizophrenia. . . ." These words are from Franz Kallmann's introduction to his classic work *The Genetics of Schizophrenia* [1]; and although they were written in 1938, they still have a certain ring of currency. Kallmann began his studies nine years earlier at the German Research Institute for Psychiatry in Munich, which had been founded shortly after the turn of the century by Ernst Rudin, who himself published one of the first papers on genetic factors in schizophrenia.[2] The aim of Kallmann's investigations was to provide "conclusive proof" that schizophrenia is an inherited disease. The pioneers in genetic studies of schizophrenia may not have realized how elusive such proof would be, for its pursuit has now spanned well over half a century, evolving into an international research effort that has produced hundreds of reports, papers, and books. Has Kallmann's goal been reached yet? Is there now proof of significant genetic factors in schizophrenia—and is it conclusive? Too often in psychiatry, as elsewhere, such judgments are made on the basis of dogma, subjective experience,

* The authors gratefully acknowledge the invaluable assistance rendered by Dr. Steven S. Matsuyama in assembling the data upon which this review is based and preparing it for visual presentation.

and wishful thinking, rather than on the basis of fact and dispassionate reason. The major purpose of this chapter is to provide an overview of the accumulated scientific evidence for the operation of genetic factors in schizophrenia by means of a relatively nontechnical summary of selected research findings. This review is by no means exhaustive but will, it is hoped, be sufficiently comprehensive to allow the reader to make his own informed judgments about proof and conclusiveness.

Research Evidence

The bulk of evidence supporting the idea of a genetic basis for schizophrenia tends to be divided among three somewhat different categories of research. The first may be termed that of family studies; in general, these examine the frequency of schizophrenia found among relatives of persons with the disorder. The second category is that of twin studies, in which the frequency of schizophrenia in monozygotic twins is compared with that in dizygotic twins. The third major category is that of adoption studies; these investigations focus on the frequency with which schizophrenia is encountered in children who have not been reared by their biologic, schizophrenic parents.

The data in the various figures and tables that follow have been condensed or abridged in order to present the essence of their significance and to avoid nonessential details. One technical point that does appear regularly is that of age-correction of frequency rates. Greater accuracy in determining morbidity risk is provided by statistical manipulation of uncorrected or crude data. This is necessary because the period of risk for the age of onset of schizophrenia is quite long, commonly held to range from ages 15 to 50. Since none of the studies is totally prospective in the sense of being complete for all individuals at risk throughout the period of risk, some estimate of an individual's average risk of becoming schizophrenic is needed. These corrections for age tend to raise the uncorrected, observed rates to more accurate approximations.

FAMILY STUDIES

The clinical observation that schizophrenia tends to "run in families" antedates the coining of the diagnostic term by Eugen Bleuler and can be traced at least to the Belgian psychiatrist B. A. Morel, who was the first to use the term dementia praecox [3] and who believed that the disorder was transmitted from one generation to another.[4] Pedigree and family studies of schizophrenia have verified this empiric fact. However, confirming that an illness tends to have a higher incidence in some family groups than in others only suggests, at most, that hereditary factors are operant. Parasitosis, hypovitaminosis, iron-deficiency anemia, pneumoconiosis, and effects of toxins are but a few examples of disease states that may occur more frequently in some families than in others, but whose familial patterns are known to result predominantly from environmental factors. Furthermore, characteristics such as language, food preferences, dress habits, behavior codes, and the like may also tend to be familial, but these are generally held to be traits that are primarily acquired rather than inborn. The point is, of course, that association alone does not establish a cause-and-effect process. If truly genetic factors are present in schizophrenia, therefore, there ought to be some correlation between the degree of genetic relationship and the frequency of the disorder in relatives of affected persons. For instance, if genetic transmission is involved in the development of schizophrenia, then parents of schizophrenics should manifest a higher frequency of the disorder than aunts and uncles, since children share a greater number of common genes with their parents than they do with more distant relatives.

The largest series involving such observations is Kallmann's twin-family study,[5] which includes determinations of the expectancy of schizophrenia among the relatives of index (affected) cases. Kallmann's findings are summarized in Table 1. These data show that the expectancy of schizophrenia in family members who are not biologically related to the index cases (spouses and step-sibs) is not much higher than the frequency of the illness in the population at large (about 1 percent). On the other hand, there is clear association between degree of consanguinity and expectancy of schizophrenia: full sibs have twice the rate of half-sibs, and monozygotic twins have a rate nearly six times that

TABLE 1

Expectancy of Schizophrenia in Family Members of Schizophrenic Index Twins *

Relationship	Number of Persons	Cases of Schizophrenia	Uncorrected Rates (%) †	Corrected Rates (%) †
Husbands and wives	254	5	2.0	2.1
Step-sibs	85	1	1.4	1.8
Half-sibs	134	4	4.5	7.0
Parents	1,191	108	9.1	9.2
Full sibs	2,741	205	10.2	14.3
Dizygotic co-twins	517	53	10.3	14.7
Monozygotic co-twins	174	120	69.0	85.8

* From Kallmann.[5] † See text.

of dizygotic twins. It is also instructive to note that the rate for full sibs is virtually identical to that for dizygotic twins, who are genetically equivalent to full sibs. The fact that the expectation for the non-consanguineous relatives exceeds that for the general population is probably related to the nuances of mate selection. There are doubtless various environmental influences, such as hospitalization, that tend to increase the opportunity for one individual with schizophrenia to meet and marry another. Indeed, while schizophrenics are more likely to remain single than are members of the population at large, approximately 2 percent (at least) of schizophrenic persons who do marry select marriage partners who are also schizophrenic.[6] This rate is twice that which would be expected in random selection of spouses.

Kallmann's data provide a good example of the effect that age correction has on expectancy rates (Table 1). The uncorrected column shows the rates for all cases of schizophrenia or suspected schizophrenia and includes all relatives over 15 years of age. The corrected column shows only definite cases of schizophrenia and is based on one-half of the relatives in the age group 15–44, plus all those over 44 years old. Thus, even when cases in which the diagnosis is in doubt are excluded, the statistical process of age adjustment reveals corrected expectancy figures that are consistently greater than the crude or observed rates.

A more comprehensive look at the world literature on family studies

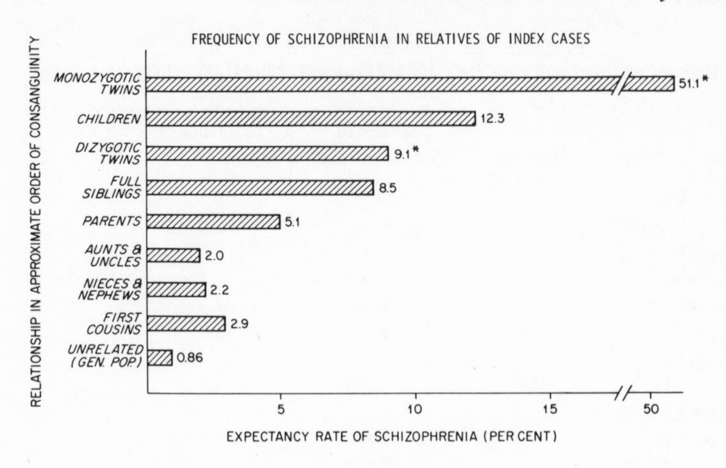

Data based on Zerbin-Rudin from Shields and Slater,[8] except for twins. Age-correction by shorter Weinberg Method, risk period 15-39 years. Twin data based on literature review of 13 studies. *No age-correction.

Fig. 1

in schizophrenia is provided in Figure 1, which is based on an extensive review by Zerbin-Rudin spanning 50 years of reports from numerous countries.[7] For comparison, Figure 1 also includes expectancy rates for twins averaged from a review of the world literature, which will be discussed in greater detail under the following section on twin studies. Once again, what is seen in these pooled data is a consistent progression of increasing expectancy of schizophrenia among relatives of index cases, which corresponds to increasing consanguinity. Stated another way, the family studies demonstrate that the greater the genetic similarity between a schizophrenic person and his relative, the greater the likelihood that the relative also has or will develop schizophrenia.

Another form of family study that many investigators have explored is that of children born to schizophrenic parents. If genetic transmission operates in schizophrenia, then it should follow that children with a schizophrenic parent should have a substantially higher frequency of schizophrenia than that found among the population at large. Further, if both parents have schizophrenia (dual matings), the frequency of the disorder among the offspring should be increased even more; and this is indeed what has been found. A summary of six studies of the offspring

TABLE 2

Frequency of Schizophrenia Reported in Children with at Least One Schizophrenic Parent *

Author	Year	Age-Corrected Risk Definite DX		(%) † All Cases
Hoffman	1921	7.0		9.4
Oppler	1932	9.7		9.7
Gengnagel	1933	8.3		8.3
Kallmann	1938	13.9		16.4
Garrone	1962		16.9	
Reisby ‡	1967	3.5		6.9
Average for All Studies: [a]		10.7		13.2

* Data based on Zerbin-Rudin [7] except for Reisby.[9]
† Age correction usually by abridged Weinberg method.
‡ Strömgren age-correction method. [a] Excludes Garrone.

of matings in which at least one of the parents was schizophrenic is presented in Table 2. The right-hand column of age-corrected frequency rates includes offspring with diagnosed or suspected schizophrenia, while the left-hand column excludes doubtful cases. (Differentiation of these categories for the study by Garrone is not clear.) Both averaged expectancy rates compare well with the rate of 12.3 percent for the children of schizophrenics noted in Figure 1; and all of the rates exceed substantially the 1 percent figure for the population at large. More striking are the results of studies of the offspring of matings in which both parents have schizophrenia. Five such studies are summarized in Table 3. It is apparent that children with two schizophrenic parents have about double the morbidity risk of those with only one affected parent. Strictly speaking, the crude rates of Table 3 are not comparable with the age-corrected rates shown in Table 2. When the combined data for dual matings are corrected for age, however, the morbidity risks rise to 39.2 percent for definite cases of schizophrenia and 44.4 percent when one-half of the doubtful cases are included.[10]

With considerable consistency, then, the various family studies of schizophrenia over the years demonstrate a clear and direct relationship between the occurrence of schizophrenia and the degree of genetic similarity throughout the full extent of the family structure. While this asso-

TABLE 3

Frequency of Schizophrenia Reported in Children of Schizophrenic Couples *

Author	Year	Number of Children	Number Definite	Schizophrenic Doubtful	Crude Rate (%) †
Kahn	1923	17	7	2	41.1 (52.9)
Kallmann	1938	35	13	3	37.1 (45.7)
Schulz	1940	59	13	5	22.0 (30.5)
Elsasser	1952	56	12	3	21.4 (26.8)
Lewis	1957	27	4	0	14.8 (14.8)
	Total	194	49	13	25.2 (31.9)

* Adapted from Erlenmeyer-Kimling [10]; not corrected for age.
† Rates in parentheses include both definite and doubtful schizophrenics.

ciation of morbidity risk with consanguinity does not establish inheritance of schizophrenia, it is certainly in accord with the hypothesis of genetic transmission. And this hypothesis is further strengthened by the evidence that has been provided by studying twins.

TWIN STUDIES

Not long after the first reports appeared on the familial epidemiology of schizophrenia, investigators began to look at twins. Researchers reasoned that, if genetic transmission plays a part in the etiology of schizophrenia, then it should become evident when concordance rates for the disorder among monozygotic twins are compared with those for dizygotic twins. The hypothesis can be simplistically dichotomized into heredity on the one hand and environment on the other. If schizophrenia results from predominantly environmental influences, then zygosity should have little or no effective relationship to the observed frequency of the disorder among twins, each pair being exposed, as it were, to roughly the same experiential circumstances. If, however, the basis for the disorder is primarily inherited, then zygosity would be of considerable importance. That is, if one member of a pair of monozygotic twins has been diagnosed as schizophrenic, then, according to the genetic transmission hypothesis, the co-twin would probably also suffer from the disorder. In other words, the concordance rate would be high because of the fact that monozygotic twins are genetically equivalent. Dizygotic twins, on the other hand, are genetically no more similar than

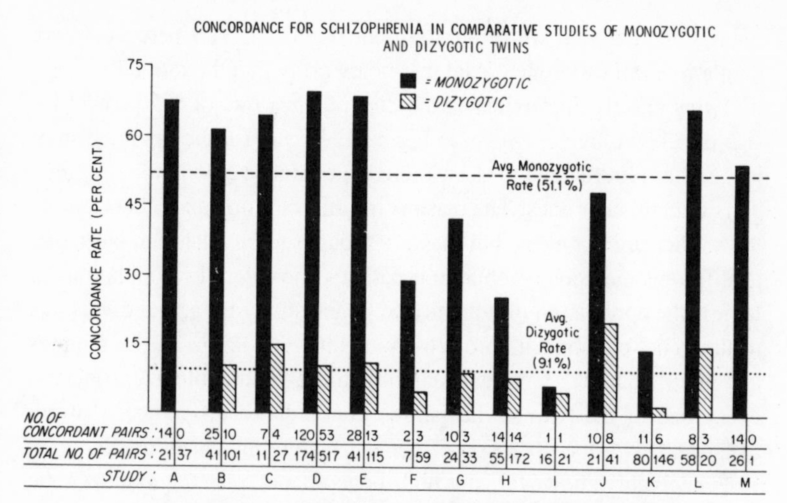

STUDY:	A		B		C		D		E		F		G		H		I		J		K		L		M	
NO. OF CONCORDANT PAIRS:	14	0	25	10	7	4	120	53	28	13	2	3	10	3	14	14	1	1	10	8	11	6	8	3	14	0
TOTAL NO. OF PAIRS:	21	37	41	101	11	27	174	517	41	115	7	59	24	33	55	172	16	21	21	41	80	146	58	20	26	1

Study = Sources of data: A-Luxemburger[11] Germany. B-Rosanoff, et al[12] USA. C-Essen-Moller[13] Sweden. D-Kallmann[5] USA. E-Slater[14] England. F-Harvald & Hauge[15] Denmark. G-Gottesman & Shields[16] England. H-Kringlen[17] Norway. I-Tienari[18,19] Finland. J-Fischer, et al[20] Denmark. K-Pollin, et al[21] USA. L-Inouye[22] Japan. M-Mitsuda[23] Japan.

Fig. 2

NOTE: CC = Number of concordant pairs; N = Total number of pairs;——Average monozygotic rate (51.1%); . . . . Average dizygotic rate (9.1%).

are nontwin full sibs. Therefore, the genetic hypothesis would predict a concordance rate for two-egg twins substantially lower than that for single-egg twins and approximately the same as that observed for nontwin full sibs of schizophrenics.

Thus far, at least 13 investigations of twin concordance rates have been reported.[5, 11–23] The results of these studies are depicted in Figure 2. What is immediately apparent from the graph is that, with one exception, concordance rates for monozygotic twins are more than double those for dizygotic twins. The exception is Tienari's study, which employed quite narrow diagnostic criteria; and even there, when the broadest concept of concordance was applied, a rate of 36 percent was calculated.[19] Nevertheless, Tienari did not have a single pair of monozygotic twins concordant for hospitalization, which is a most unusual finding. Kringlen [25] reasons that concordant pairs may have been inadvertently lost from Tienari's sample as a result of the exclusion of half of his original pairs because of the death of one or both twins. Pooling the data of all 13 studies, we find that the average concordance rate for dizygotic twins is 9.1 percent, while that for monozygotics is 51.1 percent—more than a fivefold difference. And it is noteworthy that the rate for dizygotic twins closely approximates the concordance rate of 8.5 percent for sibs of schizophrenics shown in Figure 1. Except for the Japanese studies, earlier reports tend to show more concordance among monozygotic pairs than do later ones. The reasons for this variation and others among the studies are not clear, but they are thought to be related in large part to different methods of obtaining patients (hospitals vs populations at large), the application of nonuniform diagnostic criteria, and diversities in the genetic composition or environmental conditions of the samples and their sources.[24] Notwithstanding these methodologic problems, every one of the twin studies shows concordance differences that are consistent with the operation of genetic factors in schizophrenia.

Conceivably, however, the high concordance rate for monozygotic twins might be due to nongenetic operants. Perhaps there is something about the fact of monozygosity itself that would tend to be "schizophrenogenic." This notion was refuted rather easily and well by Hoffer and Pollin [26] in their report on the overall incidence of schizophrenia among monozygotic twins. After studying nearly 16,000 veteran twin pairs, these investigators found the incidence of schizophrenia to be 1.1 percent, which is very similar to that for the general population. Moreover, the incidence among their monozygotic twins was less, but not significantly so, than among the dizygotic pairs. Thus, neither twinship nor zygosity appears to be a predisposing factor in schizophrenia.

Another proposition that has been put forth to explain the higher monozygotic concordance rates is that these twins may share so unique an experience of co-identification that the development of schizophrenia in one twin is likely to induce a similar disorder in the other. This hypothesis is difficult to test experimentally. However, its merit in contradicting a genetic basis for the higher concordance rates is weakened rather seriously by observations of monozygotic twins who have been reared apart from one another. Among the various twin studies, eight reported concordance rates for monozygotic pairs brought up separately. The results are shown in Table 4. (Isolated case reports are excluded; their meaning is difficult to interpret, since there is a tendency to report concordant but not discordant pairs.) The unusual circum-

TABLE 4
Schizophrenia in Monozygotic Twins Reared Apart *

Author	Country	Year	Age at Separation	Concordant Pairs	Discordant Pairs
Kallmann	Germany	1938	Soon after birth	1	0
Essen-Moller	Sweden	1941	7 years	1	0
Kallmann and Roth	USA	1956	Not stated	1	0
Shields	England	1962	Birth	1	0
Tienari	Finland	1963	3 and 8 years	0	2
Kringlen	Norway	1967	3 and 22 months	1	1
Mitsuda	Japan	1972	Infancy	(5)	1
				8 †	
Inouye	Japan	1972	Before 5 years	(6)	3
			Totals:	13(65%)	7(35%)

* After Slater and Cowie,[27] except for Mitsuda [23] and Inouye [28] and the omission of a single case report which was not part of a large-scale study.
† Three concordant pairs observed by both authors [29] counted only once.

stance of separate rearing from a very early age took place in only a small number of pairs, but the direction of the data is striking—concordant pairs outnumber discordant ones by more than two-to-one. Indeed, the concordance rate of 65 percent is even greater than the average rate of 51 percent found for all monozygotic twins in the worldwide literature (Figure 2). This relationship persists, moreover, when only the data of those authors common to both Figure 2 and Table 4 are included in the calculations. This reveals an average concordance rate of 56 percent for those monozygotic pairs who were reared together, compared with a rate of 65 percent for the twins reared apart. As Rosenthal [30] has observed, it is difficult to explain how such high concordance rates could have occurred in the absence of important genetic factors.

These twin studies, taken together with the data from the studies of families, provide a large and impressively consistent body of evidence in support of a genetic hypothesis in schizophrenic disorders. Even so, there always remains the possibility that the data could be explained on the basis of environmental similarities among relatives through direct or indirect association. Familial traditions, expectations, myths, rearing practices, psychodynamics, overt psychopathology, and other such psychologic forces might certainly be influential in the development of schizophrenia in relatives who, more or less, share a common environment. There is one further category of epidemiologic work, however, which obviates such arguments and may provide the most solid evidence yet available—studies of adoptees.

ADOPTION STUDIES

Ideally, adoption studies eliminate shared environmental influences, since they focus on individuals reared apart from their biologic relatives. Such investigations are relatively recent, so reports are not numerous. They are, nonetheless, consistent with what the genetic theory predicts. A summary of some of the data from three independent studies is presented in Table 5. Heston [31] studied children who were permanently separated from their schizophrenic biologic mothers within the first days of life. A carefully matched control group of adoptees with nonschizophrenic mothers was established for comparison in this retrospective study. The results are clear: None of the adoptees with unaffected mothers was found to have schizophrenia, whereas the disorder was present in 10.6 percent of the children with schizophrenic mothers.

TABLE 5
Adoption Studies of Schizophrenia *

Investigator	Year	Sample Characteristics	Sample Size	Schizophrenic	Crude Rate (%)
Heston	1966	Adopted offspring of schizophrenic mothers	47	5	10.6
		Adopted offspring of nonschizophrenic mothers	50	0	0.0
Karlsson	1966	Biologic sibs of adopted schizophrenics	29	6	20.7
		Foster sibs of adopted schizophrenics	28	0	0.0
Rosenthal et al	1971	Adopted offspring of schizophrenic parents	48(52)	3(6)	6.3(11.5)
		Adopted offspring of normal parents	67	0(2)	0.0(3.0)

* Data in parentheses include probable schizophrenia (see text).

And when the data were corrected for age, the morbidity risk rose to 16.6 percent. Significantly, the rate of 16.6 percent is higher than the corrected expectancy rates for schizophrenia among the nonadopted offspring of schizophrenics noted in Figure 1 (12.3 percent) and in Table 2 (13.2 percent).

Karlsson [32] compared the incidence of schizophrenia in the biologic sibs and foster sibs of adopted schizophrenics. None of the foster sibs was found to be affected, but one-fifth of the biologic sibs were diagnosed as having schizophrenia. This finding is again consistent with genetic transmission, but the rate is much greater than the 8.5 percent reported in the world literature for biologic sibs of schizophrenics (Figure 1). The reason for this disparity is not clear but may well be related to the fact that Karlsson's sample was from Iceland, and that essentially isolated population may have unique characteristics that render invalid any comparison of the data with those from other sources.

Rosenthal and his colleagues [33] compared 76 adopted children of parents who suffered from disorders that they classified as "schizophrenic spectrum diagnoses" with 67 adopted children of a matched control group of parents who had no history of psychiatric illness. The children in both groups had been transferred to the adoptive families at an average age of about six months. The concept of a spectrum diagnosis is crucial to a realistic understanding of these investigators' results. That is, the diagnostic categories in their schizophrenic spectrum range from schizoid personality through such descriptions as "schizophreniform borderline" to chronic or process schizophrenia. Employing this broad definition of schizophrenia, they found, not surprisingly, relatively high frequencies of spectrum diagnoses. Among the 76 offspring of parents with such disorders, 24 (31.6 percent) fell within the spectrum. Similarly, application of the spectrum diagnosis concept to the 67 adopted offspring of normal parents revealed that 12 (17.8 percent) of these children also had such disturbances. Only 48 of the adopted children in the series, however, had parents who had schizophrenia in a strict sense, and 4 additional children had parents who had either pseudoneurotic or borderline schizophrenia. If this diagnostic adjustment is applied to the children of these parents, the resulting rates for schizophrenia are 6.3 percent and 11.5 percent, respectively. Rates for the 67 adopted children of normal parents fall to zero when a strict definition of schizophrenia is used, and to 3.0 percent when cases of pseudoneurotic and borderline schizophrenia are included. After adjustment, then, the data for both experimental and control groups become similar to the frequencies noted by other investigators.

Other studies by this same group of investigators [34,35] have also shown that there is a substantially higher prevalence of schizophrenia and other psychopathology among the biologic relatives of adopted schizophrenics than among their adoptive relatives. Also of interest is their report on a newer research strategy in adoption studies called crossfostering.[36] In this special technique, three groups of adoptees are compared: offspring of normal biologic parents reared by schizophrenic adopting parents (crossfostered group); children with normal biologic parents reared by normal adopting parents (normal control group); and offspring of schizophrenic parents reared by normal adopting parents. Application of this technique to the Danish population that was the basis for this group's earlier report [33] revealed that there was a greater prevalence of psychopathology in the adopted offspring of schizophrenic biologic parents than among the normal control group or among the crossfostered group. The tentative conclusion reached from these findings was that genetic factors, rather than familial psychopathology, are important in the etiology of schizophrenia.

Methodologic Problems and the Direction of the Evidence

Perhaps the greatest shortcoming of the research evidence, in light of the many years of effort expended in study, is that it remains inconclusive: the theory of genetic transmission in schizophrenia (as well as the theories of psychologic causation) has not been proved in the strictest scientific sense. One is left with a large body of evidence from which to draw conclusions. Such inferences are certainly appropriate, if not unavoidable. Caution is warranted, however, because of numerous problems of methodology that are present in variable or unknown degrees even in the best research designs. These problems are too numerous and complex to explore at length here, but they include such things as errors in sampling techniques, biases introduced because of cultural and socioeconomic differences among the populations from which the samples are selected, and the failure to apply measures of statistical significance to the data. There are wide variations in diagnostic features; no uniform definition of schizophrenia is employed; and diagnostic criteria are often not specified. Because of the long period of risk for schizophrenia, prospective investigation tends to be incomplete, and many studies report what are actually prevalence rates at a given time. Expectancy rates can usually be estimated only by means of manipulating the data to correct for age. Such adjustments are not made universally, and reported data are often not sufficiently detailed to allow others to correct them for age. In twin studies, where the determination of zygosity is rather critical, there is often insufficient explanation of how zygosity was established. There may be overlap between different studies, resulting in the undetected inclusion of the same cases in more than one investigation. Efforts to abstract or test data reported in foreign languages are sometimes hampered by difficulties in translation. Most of the studies imply dichotomization of etiology (environment vs heredity) and of diagnosis (schizophrenia or not schizophrenia); earlier investigators in particular seemed at times to forget that such divisions are quite arbitrary and are more related to the constructs of research design than to the real world. Finally, there is the common problem of investigator bias. The genetic researchers, by and large, have a clear orientation toward, if not an investment in, the inheritance theory, and their critics tend to be equally biased in favor of psychogenic theories. It is quite probable that these several methodologic problems account in large part for the variations in findings that have been observed from study to study.

Taking these difficulties and drawbacks into consideration, what is the direction of the evidence? First, there is rather impressive consistency in the general thrust of the findings noted by many investigators over many years. Second, three different bodies of evidence—family studies, twin studies, and adoption studies—have all produced data that are in accord with what is predicted by a theory of genetic transmission.

This combined evidence argues very strongly indeed that genetic factors are operant in the etiology of schizophrenia, and it suggests further that genetic predisposition may even be a necessary condition for the development of schizophrenia. What are these "genetic factors"? How do the data fit with genetic theory? What is it that is inherited, and how is it transmitted? The answers to these questions are as difficult as the questions are important. Clearly, the complexities of the history, signs, and symptoms of the clinical syndrome subsumed under the term "schizophrenia" do not lend themselves to simple analysis, unlike some straightforward genetic traits such as blood groups. In their recent reviews of genetic research in schizophrenia, Rosenthal [30] and Zerbin-Rudin [24] note that several possible mechanisms of inheritance have been advocated, including recessiveness vs dominance; monogenic, two-gene, or polygenic action; and so forth. Many theorists currently favor the theory of inheritance through the interaction of several gene pairs (polygeny). For example, Vartanian and Gindilis [37] have observed that the risk of schizophrenia among the sibs of a schizophrenic varies in direct relationship with the number of relatives (parents, aunts, and uncles) who are psychotic. That is, if there are no such relatives, then the morbidity risk for sibs of index cases is about 10 percent. When there are three affected relatives in the parent generation, however, the risk rises progressively to 31 percent. These investigators reason that this observation can be explained only on the basis of polygenic inheritance. On the other hand, the monogenic theory advocated by Slater [38] is not without recent support, since Heston,[39] viewing schizoid and schizophrenic persons as comprising a spectrum, argues that there is evidence to substantiate a monogenic dominance hypothesis. And to complicate the picture still further, Kidd and Cavalli-Sforza [40] in a statistical analysis of summarized data concluded that, assuming a "threshold model" for the development of schizophrenia, both the polygenic and monogenic theories are in approximate agreement with observed data.

Thus, while the research data indicate that genetic factors are important, if not fundamental, in schizophrenia, they do not answer the more complex questions about the specific nature of their modes of action. Since the methodologies of the family, twin, and adoption studies appear to lack the sophistication necessary to focus with precision on these issues, the undertaking of further such investigations is probably not warranted. What is needed now are different approaches, utilizing newer technologies, and these are beginning to appear in the literature. For instance, Wyatt and his co-workers [41] have studied monoamine oxidase activity in monozygotic twins discordant for schizophrenia. They found that the activity of this enzyme in platelets was significantly lower in both schizophrenic and nonschizophrenic co-twins than in normals. Further, they noted an inverse correlation between degree of schizophrenic pathology and enzyme activity. They concluded that their data suggest that reduced monoamine oxidase activity may provide a so-called genetic marker for vulnerability to schizophrenia. It is hoped that, as a result of refinements in research strategies, our knowledge of the genetics of schizophrenia will become more complete than it is at present.

There can be no question that schizophrenia is a major medical problem; notwithstanding current therapies, it will continue to plague us until we learn much more about both its biologic and environmental determinants. In fact, it may become a still more serious problem if this crucial knowledge is not forthcoming, for the reproductive rate among schizophrenics appears to be on the increase,[42,43] and there may well be aspects of the disorder that are advantageous for survival in terms of natural selection.[6,44] (Such a mechanism, for which genetic transmission is a necessary condition, would help to explain how schizophrenia has persisted for so long at an apparently stable incidence within populations.) These observations suggest that the need for reliable methods of prediction and identification, as well as effective means of prevention and treatment, will become even more pressing in the future than they are at present. The demonstration that schizophrenia has a genetic basis might at first seem to preclude the discovery of measures to conquer it. However, the converse will more likely prove to be the case, and the presence of strong genetic influences is cause for optimism rather than discouragement. The reasons for hope can be seen in the parallel case of mental retardation. Major advancements in identification, treatment, and prevention have been made in genetically determined types of retardation, the classic example being phenylketonuria.

While advocating a predominantly psychologic orientation, Arieti [45] recently acknowledged the presence of nonpsychologic components of schizophrenia that are "most probably hereditary." Arieti opined further that ". . . the genetic claims made by such people as Rudin and Kallmann have not passed the test of time." Rudin and Kallmann were pioneers on the frontiers of medical genetics. The trails they blazed in the pursuit of knowledge were necessarily crude, and their early hypotheses and conclusions may seem rough-hewn today. While these two scientists, their contemporaries, and their successors may not have proved "conclusively" and to everyone's satisfaction that schizophrenia is an inherited disease, they have come close; and the paths they cleared are now leading to new and continuing research challenges in the genetics of schizophrenia.

REFERENCES

1. Kallmann, F. J.: *The Genetics of Schizophrenia.* New York: J. J. Augustin, 1938.

2. Rudin, E.: *Zur Vererbung und Neuentstehung der Dementia Praecox.* Berlin: Springer Verlag, 1916.

3. Kolb, L. C.: *Noyes' Modern Clinical Psychiatry,* 7th ed. Philadelphia: W. B. Saunders, 1968, p. 356.

4. Rosenthal, D.: An historical and methodological review of genetic studies in schizophrenia. In Romano, J. (ed.): *The Origins of Schizophrenia.* Proceedings of the First Rochester International Conference on Schizophrenia. Amsterdam: Excerpta Medica, 1967, pp. 15–26.

5. Kallmann, F. J.: The genetic theory of schizophrenia: an analysis of 691 schizophrenic twin index families. *Am. J. Psychiatry 103:*309–322, 1946.

6. Erlenmeyer-Kimling, L. and Paradowski, W.: Selection and schizophrenia. *American Naturalist 100:*651–665, 1966.

7. Zerbin-Rudin, E.: Endogene Psychosen Schizophrenien. In Becker, P. E. (ed.): *Humangenetik, ein Kurzes Handbuch in Fünf Banden,* Stuttgart: Thieme, 1967, vol. 2, pp. 446–504.

8. Shields, J. and Slater, E.: Genetic aspects of schizophrenia. *Hosp. Med. 1:*579–584, 1967.

9. Reisby, N.: Psychosis in children of schizophrenic mothers. *Acta Psychiatr. Scand. 43:*8–20, 1967.

10. Erlenmeyer-Kimling, L.: Studies on the offspring of two schizophrenic parents. *J. Psychiatr. Res.* (*suppl. 1*) *6:*65–83, 1968.

11. Luxemburger, H.: Psychiatrisch-neurologische Zwillings-pathologie. *Zentralbl. Gesamte Neurol. Psychiatr. 56:*145, 1930.

12. Rosanoff, A. J., Handy, L. M., Plesset, I. R., et al: The etiology of so-called schizophrenic psychoses, with special reference to their occurrence in twins. *Am. J. Psychiatry 91:*247–286, 1934.

13. Essen-Moller, E.: Psychiatrische Untersuchungen an einer Serie von Zwillingen. *Acta Psychiatr. Neurol. Suppl. 23:*1–200, 1941.

14. Slater, E.: *Psychotic And Neurotic Illnesses in Twins.* Medical Research Council Special Report Series no. 278. London: H.M. Stationery Office, 1953.

15. Harvald, B. and Hauge, M.: Hereditary factors elucidated by twin studies. In Neel, J. V., Shaw, M. W., and Schull, W. J. (eds.): *Genetics and the Epidemiology of Chronic Diseases.* Washington, D.C.: Department of Health, Education, and Welfare, 1965, pp. 61–76.

16. Gottesman, I. I. and Shields, J.: Schizophrenia in twins: 16 years' consecutive admissions to a psychiatric clinic. *Br. J. Psychiatry 112:*809–818, 1966.

17. Kringlen, E.: *Heredity and Environment.* In *The Functional Psychoses: An Epidemiological Clinical Twin Study.* London: Heineman, 1967.

18. Tienari, P.: Psychiatric illness in identical twins. *Acta Psychiatr. Scand.* (*suppl. 171*) *39:*1–195, 1963.

19. Tienari, P.: Schizophrenia in monozygotic male twins. *J. Psychiatr. Res.* (*suppl. 1*) *6:*27–36, 1968.

20. Fischer, M., Harvald, B., and Hauge, M.: A Danish twin study of schizophrenia. *Br. J. Psychiatry 115:*981–990, 1969.

21. Pollin, W., Allen, M. G., Hoffer, A., et al: Psychopathology in 15,909 pairs of veteran twins: Evidence for a genetic factor in the pathogenesis of schizophrenia and its relative absence in psychoneurosis. *Am. J. Psychiatry 126:*597–610, 1969.

22. Inouye, E.: A search for a research framework of schizophrenia in twins and chromosomes. In Kaplan, A. R. (ed.): *Genetic Factors in "Schizophrenia."* Springfield: Charles C. Thomas, 1972, pp. 495–503.

23. Mitsuda, H.: Heterogeneity of schizophrenia. Ibid., pp. 276–293.

24. Zerbin-Rudin, E.: Genetic research and the theory of schizophrenia. *Int. J. Ment. Health 1:*42–62, 1972.

25. Kringlen, E.: Schizophrenia in twins: An epidemiological-clinical study. *Schizophrenia Bulletin.* Bethesda: National Institute of Mental Health, Dec. 1969, pp. 27–39.

26. Hoffer, A. and Pollin, W.: Schizophrenia in the NAS-NRC panel of 15,909 veteran twin pairs. *Arch. Gen. Psychiatry 23:*469–477, 1970.

27. Slater, E. and Cowie, V.: *The Genetics of Mental Disorders.* London: Oxford University Press, 1971, p. 41.

28. Inouye, E.: Monozygotic twins with schizophrenia reared apart in infancy. *Jap. J. Hum. Genet. 16:*182–190, 1972.

29. Inouye, E.: Personal communication, 1973.

30. Rosenthal, D.: Genetic research in the schizophrenic syndrome. In Cancro, R. (ed.): *The Schizophrenic Reactions: A Critique of the Concept, Hospital Treatment, and Current Research.* New York: Brunner/Mazel, 1970, pp. 245–258.

31. Heston, L. L.: Psychiatric disorders in foster home reared children of schizophrenic mothers. *Br. J. Psychiatry 112:*819–825, 1966.

32. Karlsson, J. L.: *The Biologic Basis of Schizophrenia.* Springfield: Charles C. Thomas, 1966, p. 79.

33. Rosenthal, D., Wender, P.H., Kety, S. S., et al.: The adopted-away offspring of schizophrenics. *Am. J. Psychiatry 128:*307–311, 1971.

34. Wender, P. H., Rosenthal, D., Zahn, T. P., et al.: The psychiatric adjustment of the adopting parents of schizophrenics. *Am. J. Psychiatr. 127:*1013–1018, 1971.

35. Kety, S. S., Rosenthal, D., Wender, P. H., et al.: Mental illness in the biological and adoptive families of adopted schizophrenics. *Am. J. Psychiatry 128:*302–306, 1971.

36. Wender, P. H., Rosenthal, D., Kety, S. S., et al.: Crossfostering: A research strategy for clarifying the role of genetic and experiential factors in the etiology of schizophrenia. *Arch. Gen. Psychiatry 30:*121–128, 1974.

37. Vartanian, M. Y. and Gindilis, V. M.: Some notes on the genetics of behavior traits in man, especially of abnormal mental traits like schizophrenia. *Soc. Biol. 20:*246–253, 1973.

38. Slater, E.: The monogenic theory of schizophrenia. *Acta Genet. Statist. Med. 8:*50–56, 1958.

39. Heston, L. L.: The genetics of schizophrenic and schizoid disease. *Science 167:*249–256, 1970.

40. Kidd, K. K. and Cavalli-Sforza, L. L.: An analysis of the genetics of schizophrenia. *Soc. Biol. 20:*254–265, 1973.

41. Wyatt, R. J., Murphy, D. L., Belmaker, R., et al.: Reduced monoamine oxidase activity in platelets: A possible genetic marker for vulnerability to schizophrenia. *Science 179:*916–918, 1973.

42. Kallmann, F. J., Falek, A., Hurzeler, M., et al.: The developmental aspects of children with two schizophrenic parents. *Psychiatric Research Report 19.* Washington, D.C.: American Psychiatric Association, 1964, pp. 136–145.

43. Erlenmeyer-Kimling, L., Nicol, S., Rainer, J. D., et al.: Changes in fertility rates of schizophrenic patients in New York state. *Am. J. Psychiatry 125:*916–927, 1969.

44. Jarvik, L. and Chadwick, S. B.: Schizophrenia and survival. In Hammer, M., Salzinger, K., and Sutton, S. (eds.): *Psychopathology.* New York: Wiley, 1972, pp. 57–73.

45. Arieti, S.: An overview of schizophrenia from a predominantly psychological approach. *Am. J. Psychiatry 131:*241–249, 1974.

6

GENETICS OF MENTAL RETARDATION *

Barbara F. Crandall, M.D. and
George Tarjan, M.D.

Introduction

In current clinical practice, a valid diagnosis of mental retardation requires that four criteria be met: (1) significantly impaired intelligence (an IQ of 70 or less on psychometric tests), (2) similarly impaired general adaptation, (3) the concurrent presence of these two cardinal symptoms, and (4) onset prior to 17 years of age. There are approximately 6 million individuals (3 percent of the population) in the United States who have IQs of less than 70, but of these, some 4 million do not show sufficient impairment of general adaptation to satisfy the second requirement. Hence there are approximately 2 million retardates, a rate of <1 percent.

Mental retardation is an extremely heterogeneous condition. The 2 million mentally retarded can be divided into two fairly distinct groups, with 75 percent constituting the "sociocultural" or "psychosocial" category. Most of these individuals are of school age and are identified

* In preparing this chapter, the authors were supported in part by HEW Grants nos. HD–04612, HD–05615, HD–00345, MH–10473, MCT–927, SRS 59–P–45192/9, and the Stanley W. Wright Memorial Fund, UCLA.

upon entering school; they are reabsorbed into the general population after leaving school. They exhibit mild degrees of retardation (IQ 50–70), with no physical stigmata or identifiable pathology, and with morbidity and mortality rates similar to those of the general population. They usually have a parent or sib of less than average intelligence. The condition is highly social-class dependent, with the economically, educationally, and socially underprivileged strata of the population being grossly overrepresented.

The role of genetic forces in causation is unclear. This large class may prove to be composed of several subgroups with varying genetic contributions. We will omit this group from our discussion and limit consideration to the genetic aspects of mental retardation in the smaller category usually described as the "clinical" types, in which the genetic etiology is more clearly identified.

Categories of "Clinical" Types of Mental Retardation

The group of retardates to be considered in this paper consists of approximately half a million individuals in the United States. They are generally moderately retarded (IQ < 50), evidence clinical or laboratory pathology, and are usually identified prior to school age. The etiology remains unknown in about half the cases. Of the remainder, approximately 25 percent are affected by a genetically related disorder, another 20 percent by pre- and postnatal infections, and 5 percent by trauma and prematurity.

No attempt will be made to encompass all the many rare genetic disorders which may result in mental retardation. Rather, some of the different categories of genetic disorders will be presented in the following subdivisions.[1–4]

Chromosome abnormalities. Seventy-five percent of the genetic types of mental retardation are due to chromosomal abnormalities. Only a small percentage—about 20 percent—are caused by single-gene disorders, and about 5 percent by malformations of the central nervous system.

Single-gene mutations. Disorders due to single-gene mutations may result in metabolic errors, often with few malformations, or in structural defects of various systems. Metabolic errors may: (1) occur without the accumulation of products in cells; (2) lead to one of the storage diseases; or (3) result from a defective transport mechanism. Some of the commoner examples of these three different mechanisms will be presented.

Malformations resulting from a combination of genetic and environmental factors. Very few such conditions lead to mental retardation, but a brief comment on the commoner central nervous system malformations will be presented. Information on genetic counseling, prevention, and treatment is included with each group of disorders. A general discussion of amniocentesis, together with some consideration of future needs in medical genetics, appears in the final section.

Chromosome Abnormalities

About 4 percent of all conceptions have chromosome abnormalities. Since many of these abnormal fetuses are aborted spontaneously,[5] only 1 percent of all newborns have chromosome abnormalities.[6]. While chromosome abnormalities affecting the autosomes (or nonsex chromosomes) result in moderate to profound mental retardation, usually with associated physical malformations, sex chromosome abnormalities are tolerated with little, if any, mental retardation. Some of the commoner chromosomal syndromes result from an additional chromosome (trisomies 21, 18, and 13), and others from loss of material from a specific chromosome. Complete absence of an autosome is usually not compatible with life. Previously, the difficulty in identifying each chromosome led to problems in associating a specific chromosome abnormality with a specific phenotype. However, the advent of the new differential chromosome stains has made it possible to identify each chromosome accurately and to detect small duplications and deletions. We can expect that previously undetected chromosome abnormalities will explain a small proportion of hitherto unspecified cases of mental retardation. Although no single phenotypic change is diagnostic of a given chromosome abnormality, some of the collective changes are sufficiently typical to suggest a specific chromosome abnormality.

TRISOMY 21 (DOWN SYNDROME)

Trisomy 21 is one of the commonest malformation syndromes known. It is found in 15–20 percent of institutionalized patients.

The more usual clinical findings in Down syndrome are well known. They include hypotonia, brachycephaly, obliquity of the palpebral fissures, epicanthal folds, abnormal and low-set ears, short neck, large furrowed tongue, simian lines, abnormal dermatoglyphics, and fifth-finger clinodactyly. No single sign is pathognomonic of the malformation; the clinical impression results from the presence of several of these findings. Recognition in the newborn period is important, and Table 1 lists the ten signs most often found in newborns. Hall noted that at least four of these were present in all newborns with Down syndrome, and six of them in 90 percent.[7] Mental retardation, although highly variable, appears to be present in every case, and the majority show a moderate degree of retardation (IQ 30–50). In the severely retarded, mental development may cease, or appear to deteriorate, after the age of 10 or 11 years. Of the higher-grade retardates, 16 percent reach a peak between 11 and 15 years, 35 percent at age 18, and 48 percent after the age 20.[8] Approximately 40 percent of Down syndrome patients have a cardiac

TABLE 1

Down Syndrome: Clinical Findings in Newborns

Finding	Percentage
1. Hypotonia	80
2. Poor Moro reflex	85
3. Joint hyperflexibility	80
4. Nuchal skin folds	80
5. Flat facies	90
6. Slanted palpebral fissures	80
7. Abnormal ears	60
8. Dysplasia of pelvis	70
9. Dysplastic middle phalanx, fifth finger	60
10. Simian crease	45

malformation, of which ventricular and atrial septal defects are the most common, followed by patent ductus arteriosus and atrioventricularis communis.[9] Other malformations sometimes found in Down syndrome include duodenal atresia, tracheoesophageal fistula, exomphalos, and Hirschsprung disease.

Down syndrome occurs once in 600 births; while the incidence is 1/2,000 births when the maternal age is 20 years, it increases to 1/40 by age 45 years. While the majority (94 percent) of patients with Down syndrome have trisomy 21, 2.4 percent are mosaics, and 3.6 percent have translocations.[10] Primary trisomy probably results from nondisjunction of chromosomes 21, usually at the first meiotic division. The maternal age relationship suggests that the nondisjunction occurred in the mother. The mechanism for this is not known, but studies in rodents have shown reduced chiasma counts in "older" eggs, and this may lead to irregular distribution of homologous chromosomes in meiosis. It is not known whether these abnormal eggs were defective initially, causing them to be ovulated and fertilized late in the reproductive life of the individual (as suggested by Edwards and Fowler [11] in the "production line" concept), or whether nondisjunction resulted from late ovulation in a previously normal egg. Some cases of trisomy 21 result from mosaicism in either parent, the trisomic cells being too limited in number or distribution to produce phenotypic changes. The presence of thyroid antibodies in a greater proportion of young mothers (under 35 years) with a Down syndrome child than in controls suggests that an immunologic mechanism might be an etiologic factor in nondisjunction.[12]

There have been several studies concerning mortality in Down syndrome. Carter found that of 700 patients with Down syndrome, 30 percent died within 1 month of birth, 53 percent by 1 year, and 60 percent by 10 years.[13] The highest mortality was between the ages of 1 and 5 years, and after 40 years [14]; between 5 and 40 years it was little above normal. Life expectancy figures in Down syndrome vary according to the group studied, but one study on noninstitutionalized cases reported this as 16.2 years at birth, 22.4 years at 1 year, 26.7 years at 5 to 9 years, and 2.5 years at 50 to 54 years of age.[15] Females had a higher mortality than males. Collmann and Stoller found that the infant mortality rate (deaths under 1 year per 1,000) was 311 for Down syndrome, as compared with 20.8 in the general population.[16] The commonest cause of death in the neonatal period was congenital heart disease, and respiratory infections accounted for the majority by the end of the first year.[10] The increased mortality from acute leukemia in children under 2 years of age with Down syndrome is between 15 and 20 times that of the general population (approximately 1 percent). In Down syndrome the propensity toward early aging and the increased incidence of Alzheimer presenile dementia are well known, and these probably account for the increased mortality after age 40. All patients with Down syndrome may show the neuropathologic changes of Alzheimer disease after 35 years of age.[17]

If possible, the diagnosis of Down syndrome should be made at birth or shortly thereafter. A routine chromosome analysis is important in every case, both to confirm the diagnosis and to identify those with translocations. If trisomy 21 is found in the propositus, we do not arrange for parental chromosome studies, unless other cases of Down syndrome have been reported in the family. It is true that chromosomal mosaics and minor variants of etiologic importance could go undetected, but we believe that the chance of finding these is very small. If one child is born with trisomy 21, the risk for another is approximately 1 percent,[18] and amniocentesis is recommended in all succeeding pregnancies. When translocation Down syndrome is detected in a patient, chromosome analysis is indicated in both parents. Translocations account for 9 percent of Down syndrome cases when the mother is less than 30 years old.[19] Approximately one-third of translocations are inherited, and the remainder arise de novo in the patient. When one parent is found to be a translocation carrier, the carrier's sibs and children are studied in order of reproductive risk. The observed risk for a second child with Down syndrome is 10–15 percent when the mother carries the translocation and 4 percent when the father carries it.[20] Amniocentesis is thus strongly indicated if either parent carries a translocation.

TRISOMY 18 (EDWARDS SYNDROME)

Trisomy 18 abnormality was first described in 1960 by Edwards et al.[21] The incidence is 1/3,000 newborns, and a number are spontaneously aborted early in pregnancy.[5] The sex ratio at birth favors females

TABLE 2
Findings Present in 50% of Patients with Trisomy 18

1. Growth deficiency
2. Mental retardation
3. Hypertonicity
4. Prominent occiput
5. Low-set, rotated, malformed ears
6. Small palpebral fissures
7. Micrognathism
8. Clenched fist, with second finger over third
9. Digital pattern of arches or low-ridge count
10. Short sternum
11. Narrow pelvis
12. Ventricular septal defect

by three-to-one. A large group of abnormalities have been ascribed to trisomy 18, but Table 2 lists only those present in 50 percent or more of the chromosomally proven cases.[22] All are severely retarded, and 75 percent die within six months after birth. While the majority of these children have an additional chromosome 18, a few have partial trisomies (resulting from translocations) or mosaicism. For this reason, as well as for diagnostic confirmation, all suspected trisomy 18 patients should have a routine chromosome analysis. Parental chromosome studies are indicated, particularly if a translocation is detected. Although most cases of trisomy 18 are sporadic, a second chromosome abnormality in a succeeding pregnancy has prompted us to recommend amniocentesis to these mothers.[23]

TRISOMY 13 (PATAU SYNDROME)

This chromosome abnormality occurs once in 5,000 births. The findings noted in 50 percent or more cases are shown in Table 3.[24] Approximately 75 percent of these patients die by six months of age, and the rest are profoundly retarded. Most commonly, chromosome analysis reveals an additional D-group chromosome, identified as a no. 13 by differential staining studies. A few patients have only 46 chromosomes, with a sporadic translocation involving two chromosomes 13, or a no. 13 and another chromosome; a few others have an inherited translocation. D-D

TABLE 3
Findings Present in 50% of Patients with Trisomy 13

1. Polydactyly
2. Congenital heart defect
3. Microphthalmia ± coloboma
4. Simian line
5. Cleft lip and palate
6. Apneic spells
7. Hyperconvex nails
8. Low-set ears ± malformation
9. Scalp defects
10. Skin folds around neck

translocations are the commonest chromosomal rearrangement in the general population (1/1,000) but have rarely resulted in a child with trisomy 13.

TRISOMY 8

Trisomy 8 was identified by Caspersson et al.[25] using differential chromosome stains, and the phenotypic changes are characteristic enough to form a recognizable syndrome. Two patients with this syndrome (both mosaics) have been seen at UCLA, and the physical findings in both were sufficiently specific to allow the diagnosis to be made prior to the chromosome study.[26] Two of Caspersson's patients were mosaics. Clinical findings include mental retardation (usually moderate), strabismus, prominent ears, vertebral anomalies, absent patellas, and abnormalities of the hands and feet. One of Caspersson's patients with mosaicism was studied, although she had no reported physical or mental defects, because she had had two spontaneous abortions. Trisomy 8 appears to carry the largest reduplicated chromosome that is compatible with survival.

PARTIAL TRISOMY 15

Another trisomy, a partial one of chromosome 15, is of interest because both the children we have described with this anomaly have no physical malformations.[27] They are of normal height, but both are moderately retarded (IQ 40–50) and very hyperkinetic. The additional no. 15 (approximately two-thirds of this chromosome) was present in all cells examined. The absence of malformations with this autosomal abnormality has caused us to revise our indications for chromosome studies, and they are now part of the work-up of all children with moderate or more severe retardation.

CHROMOSOME DELETION SYNDROMES

In addition to those involving an extra chromosome, a number of syndromes are associated with partial loss of an autosome. Some of the clinical changes associated with loss of material from chromosomes 4 and 5 are shown in Table 4.

TABLE 4

Findings in Chromosome Deletion Syndromes

Short arm of chromosome 4 4p –	Short arm of chromosome 5 5p –
1. Mental retardation—severe	1. Mental retardation—moderate
2. Failure to thrive	2. Catlike cry
3. Seizures	3. Large cornea
4. Cleft lip and/or palate	4. Antimongoloid slant
5. Depressed angles of mouth	5. Round face

Deletions commonly result from a balanced translocation in either parent. Differential stains are necessary to identify both the chromosome abnormality in the child and the translocation in the parent.

SEX CHROMOSOME ABNORMALITIES

Newborn studies indicate that the incidence of sex chromosome anomalies is about 0.25 percent, but this probably is an underestimate, as the one- or two-cell counts commonly used in such studies may fail to detect mosaic chromosome complements.

TURNER SYNDROME

Although Turner first described a syndrome of short stature, sexual infantilism, webbed neck, and cubitus valgus in 1938,[28] the chromosome abnormality, usually the loss of one X chromosome, was not reported until 1960.[29] Gonadal dysgenesis, varying from hypoplasia to complete absence of germinal elements, was found in over 90 percent of patients. The incidence of this abnormality is only 1/3,000 female births, for a large number are aborted, particularly in the second and third trimester, and it has been estimated that only 1/40 XO fetuses survive to term.[30] Most infants with Turner syndrome have a birthweight of less than 2,500 g; although some are noted to have webbed necks and edema of the dorsum of the hands and feet, the majority are ascertained because of short stature (the only consistent physical finding) and failure to develop secondary sex characteristics. Twenty percent have cardiac abnormalities (of which the most frequent is a coarctation), 50 percent have perceptive hearing losses (particularly likely to develop in adolescence), and 60 percent have renal and urinary tract abnormalities. Probably less than 10 percent are retarded, and their retardation is usually minimal. Learning problems resulting from difficulties in space-form recognition are particularly common in Turner syndrome,[31] and early identification of these problems may prevent educational difficulties later on.

The commonest chromosome finding in Turner syndrome is the loss of an X chromosome (45,X). However, about 30 percent of patients with the typical Turner syndrome phenotype have positive sex-chromatin studies resulting from mosaicism, a structural X chromosome abnormality, a deletion or isochromosome, or an XX pattern.[32] Family studies using the Xg blood group have shown that in 74 percent of patients the single X chromosome is derived from the mother.[33] This finding is consistent with the lack of maternal age effect.

KLINEFELTER SYNDROME: 47, XXY

Klinefelter syndrome describes a male phenotype with one or more additional X chromosomes (see Chapter 1). Hypogonadism is the only consistent physical abnormality, so that most patients are ascertained after puberty or in "fertility" clinics. Tall stature with long limbs and gynecomastia in 30 percent of the cases may be concomitant features. There is a greater incidence of Klinefelter syndrome in institutions for the retarded than in the public at large; approximately 25 percent are retarded, usually minimally.

This sex chromosome abnormality occurs about 1/500 births. From

Xg blood group studies, it appears that the additional X chromosome is maternally derived in 64 percent of cases.[33] A number of families have been reported in which one child has a sex chromosome abnormality (most often Klinefelter syndrome) and a second has Down syndrome. Occasionally, double aneuploidies have been described in one individual (most often Klinefelter and Down syndromes).[34] In rare instances, Klinefelter syndrome has been reported with up to five X chromosomes. Nearly all of these individuals are retarded, and both the retardation and the physical malformations increase in severity with the increasing number of X chromosomes.

The XYY and XXX syndromes are discussed in Chapter 1.

Single-Gene Mutations Causing Mental Retardation

Single-gene mutations follow mendelian patterns of inheritance. Before genetic risks can be analyzed, a specific diagnosis must be obtained, together with a detailed family pedigree. Traits manifested when only one of a pair of genes is mutant are dominant; traits not carried on the X chromosome are autosomal. Autosomal dominant disorders frequently affect several generations of a family, and only the children of an affected individual are at risk (50 percent). Late onset and variable expression of the trait are characteristic of dominant diseases. Autosomal recessive traits affect several members of a sibship and are the result of a pair of mutant genes, both parents being obligate carriers. The risk for another affected child is 25 percent; however, an affected individual or an unaffected sib who marries a normal individual rarely has affected children. Sex-linked traits are carried only on an X chromosome, as the Y appears to contribute sex determining genes only. Father-to-son transmission excludes X linkage. X-linked dominant diseases are transmitted to 50 percent of the offspring of an affected woman and to all the daughters of an affected man. X-linked recessive traits only affect males. All the daughters of an affected male are carriers, but none of his sons is at risk. If the mother is a carrier, half of her sons will be affected and half of her daughters will be carriers. Where the mother is a known carrier and there is no specific biochemical test for the enzyme in amniotic cells, the disease can be prevented by identifying males through amniocentesis, with therapeutic abortion if desired.

METABOLIC ERRORS WITHOUT STORAGE

These diseases are usually inherited as autosomal or X-linked recessive traits, and the resulting block in the metabolic pathway results in accumulation of the substrate prior to the block (as in PKU), failure to produce the end product (as in hypothyroidism), or diversion of products to alternate pathways. The defect affects small circulating molecules, so that "storage" does not occur. The largest number of metabolic errors in this category involve amino acids, and phenylketonuria (PKU) is the commonest of these.

PKU. This disease,[35] transmitted as an autosomal recessive, occurs once in 10,000–13,000 births. Many states now have compulsory screening of all newborns for this disorder. Phenylalanine hydroxylase, the enzyme responsible, is almost completely confined to the liver. Untreated children are frequently, but not always, retarded. Symptoms include an unusual odor, hyperkinetic behavior (sometimes with autistic features), and seizures. A low phenylalanine diet is necessary to prevent mental retardation and is usually continued until at least five years of age.

Women who either were never identified as having PKU or who were treated but never followed up adequately are now producing children with mental retardation. The high phenylalanine level, while having no effect on the mother, severely compromises the developing fetal brain. It now appears that 100 percent of the children of PKU mothers who were not treated during their pregnancies are retarded, and about 25 percent have congenital malformations.[36] This speaks strongly for the continuation of a low phenylalanine diet in individuals with PKU and for routine urinary screening of all pregnant women, mothers with more than one retarded child, or even children with learning problems. As phenylalanine hydroxylase is not present in amniotic fluid cells, PKU cannot be identified through amniocentesis.

Galactosemia. The commonest metabolic disorder affecting carbohydrates is galactosemia, and prompt elimination of lactose from the diet, if accomplished early enough, can prevent the symptoms from developing. Untreated cases show mental retardation, cataracts, cirrhosis, seizures, and early death. The disorder is inherited as an autosomal recessive, and affected fetuses can be identified through amniocentesis.

Cretinism. One form of athyreotic cretinism appears to be genetic and is inherited as an autosomal recessive trait. In addition, there are at least five distinct enzymatic defects resulting in failure to form thyroid hormone and inherited as autosomal recessive traits.

METABOLIC ERRORS WITH STORAGE

These diseases are the result of a defective enzyme in the catabolic pathway of a compound, leading to accumulation of the substrate in the cell.[37] This group of diseases can be further subdivided according to the class of compounds affected.

Sphingolipids. This large group of compounds is derived from a base, sphingosine, with the addition of a long-chain fatty acid and phosphorylcholine (sphingomyelin), glucose, or galactose.

(1) Tay-Sachs disease. In this disease, abnormal or absent hexosaminidase-A leads to the accumulation of gangliosides in the central nervous system. Usually a normally developing baby of six months starts to regress in development, with loss of voluntary movement, blindness, spasticity, and death by three to four years. Tay-Sachs disease is particularly common among Ashkenazic Jews, of whom 1/30 are carriers (prevalence in the general population is 1/300). Carriers can be identified by a simple serum hexosaminidase estimation. This autosomal recessive disease can be diagnosed prenatally by amniocentesis, which is recommended if both parents are carriers and are anxious to have children. Less common are the late infantile and juvenile forms of amaurotic family idiocy.

(2) Metachromatic leukodystrophy: In this condition, sulfatide accumulates in the central nervous system, kidneys, and gall bladder. In the commoner late infantile form, ataxia, impaired motor function, and hypotonia develop in the second year, with spasticity, severe mental retardation, and death by five years of age. This autosomal recessive disorder can be confirmed by the absence of arylsulphatase-A from the urine and fibroblasts. An affected fetus can be identified prenatally through amniocentesis.

(3) Gaucher disease: Only the infantile type of this disease results in mental retardation. Cerebroside accumulates in the central nervous system and liver with opisthotonos, increasing rigidity, and death by one year. This disorder is also transmitted as an autosomal recessive. Prenatal diagnosis can identify the affected fetus.

(4) Niemann-Pick disease: Sphingomyelin accumulates in the reticuloendothelial and nervous systems and in the liver. A number of clinical forms of this disease have been described, but the most common leads to failure to thrive, with blindness, hepatosplenomegaly, and death by three years of age. The visceral form spares the nervous system, but a third variety causes central nervous system deterioration in late infancy, with death by four to six years.

Mucolipids. The mucolipidoses are a recently described group of disorders originally thought to be variants of Hurler syndrome but without excretion of mucopolysaccharides in the urine.[38] The specific enzyme defect is not known, but all the lysosomal enzymes are depleted, and coarse granules are seen in fibroblasts. Three mucolipidoses have been identified with varying clinical courses, mental retardation, coarse features, gingival hypertrophy, and skeletal anomalies. These diseases are all inherited as autosomal recessives and can probably be identified prenatally.

Mucopolysaccharides. This large group of disorders includes the Hurler, Hunter, and Sanfilippo syndromes, all of which cause mental retardation. While both Hurler and Sanfilippo syndromes are transmitted as autosomal recessives, Hunter is an X-linked disorder. Mucopolysaccharides are excreted in the urine; mental retardation, coarse facies, large head, hernias, corneal clouding, short stature, hepatosplenomegaly, and joint limitation usually develop. Prenatal diagnosis can be made in Hurler and Hunter syndromes.

Polysaccharides. Pompe disease, one of the glycogen storage diseases, is associated with mental retardation. Deposition of glycogen occurs in the central nervous system and heart, with death by two years of age. The disease is inherited as an autosomal recessive, and the absence of the enzyme (alpha-1-4-glucosidase) can now be detected in amniotic fluid cells.

DEFECTS OF TRANSPORT MECHANISMS

Defects in transport mechanisms may block the absorption of substances, prevent their transport to target organs, and affect the excretion of specific compounds.

Menke syndrome. This syndrome is probably due to a defect in copper transport across the intestinal cell wall. It results in abnormal hair, failure to thrive, seizures, and spasticity. It is transmitted as an X-linked recessive.

Hartnup disease. This autosomal recessive disorder is attributed to a defect in the transport of specific neutral amino acids. A pellagra-like rash, cerebellar ataxia, and mental retardation usually respond well to nicotinamide therapy.

SINGLE-GENE MUTATIONS RESULTING IN STRUCTURAL ABNORMALITIES AND CAUSING MENTAL RETARDATION

A group of diseases with neurocutaneous abnormalities are responsible for approximately 1 percent of cases of mental retardation. They are sometimes referred to collectively as the phakomatoses.

Sturge-Weber syndrome. Hemangiomas of the face, eye, and meninges lead to seizures and mental retardation. The condition is not genetic.

Tuberous sclerosis. Of this disease group, tuberous sclerosis is the one most often seen in the institutionalized mentally retarded. Berg and Crome reported that of 2,000 institutionalized children, 20 had phakomatoses, and 16 of those were diagnosed as having tuberous sclerosis.[39] The disease is extremely variable in expression, so that these figures probably underestimate its frequency. Infantile spasms followed by myoclonic epilepsy may be the first signs of the disease. Mental retardation may be absent, mild, or moderate; however, all those with retardation develop siezures. The presence of vitiliginous spots, which fluoresce with a Wood's lamp, are helpful in making the diagnosis. The typical butterfly facial rash is frequently angiomatous at first and later becomes fibroangiomatous. Intracranial calcification, evident on the skull films, is not usually seen until after the age of ten years. Fifty percent of patients have retinal phakomas, and subungual fibromas are common. Renal tumors are seen in 80–100 percent of patients, and cystic change in the phalanges in 60 percent. Five percent of patients develop cardiac rhabdomyomas, and 5–10 percent die of brain tumors.

Tuberous sclerosis is transmitted as an autosomal dominant disorder, but nearly one-third of cases appear to be due to new mutations. If only one child is affected, both parents and sibs are examined for any evidence of disease. If family members appear to be normal, the affected child probably represents a new mutation, and there is a low risk that other children will be affected. However, should the affected individual reproduce, his children have a 50 percent risk of developing the disease.

Multiple neurofibromatosis. This is a common disease with an incidence of approximately 1/3,000. About 30 percent of patients are mentally retarded, and the retardation is usually minimal. While the expressivity is very variable, reliable positive evidence for the disease is the presence of six or more café-au-lait spots of 1.5 cm or more in their longest diameter.[40] The disease is inherited as an autosomal dominant, but at least one-third of cases appear to represent new mutations.

MISCELLANEOUS CONDITIONS

There are many syndromes of mental retardation with malformations which are transmitted as autosomal recessive traits. If the first child in a family exhibits any of these disorders, the genetic counselor faces a difficult problem. He will find it easier to identify the genetic mechanism if it resembles a known or published syndrome. If the findings do not conform to any known syndrome, if there is no consanguinity, and if the family history is negative, the advice given is that this probably represents a sporadic disorder with a low recurrence risk (5 percent).

CNS Malformations Due to a Combination of Genetic and Environmental Causes

The malformations most likely to be associated with mental retardation are spina bifida, encephalocele, and hydrocephalus. Approximately half of the children with these disorders are retarded. Hydrocephalus is frequently associated with spina bifida, and if the latter is a single malformation, the risk for another child with either spina bifida or anencephaly is 5 percent; after two, 15 percent.[41] Amniocentesis at 16 weeks gestation to determine the level of α-fetoprotein in the amniotic fluid, will probably detect 90 percent of fetuses with spina bifida or anencephaly. This test is recommended for all couples who have had a child with these malformations of the central nervous system. A sex-linked type of hydrocephalus caused by aqueductal stenosis, which may be associated with abnormal thumbs, accounts for 2 percent of hydrocephaly.[42] Otherwise, hydrocephalus as a single malformation is usually sporadic and carries a low recurrence risk.

Future Prospects

We are only beginning to understand the extent of the genetic contribution to mental retardation, and we can expect major technologic advances in diagnosis as well as prevention and treatment. In addition, better delineation of the genetic contribution to the heterogeneous group of psychosocial and familial types of retardation should lead to the adoption of more specific preventive measures.

Amniocentesis has become one of the genetic counselor's most powerful tools, in a negative as well as a positive way. It converts recurrence risk figures from a level of approximately 10 percent, 25 percent, or 50 percent to 0 or 100 percent. For example, a 25 percent risk for a second child with Tay-Sachs would probably deter a couple from undertaking another pregnancy, but amniotic cell studies can enable them to complete only those pregnancies in which a normal child has been identified. At present, amniocentesis is usually recommended where there is a risk of 1 percent or more for a child with a chromosome abnormality (usually Down syndrome), for sex in a sex-linked disorder, and for a metabolic disorder where the abnormality can be detected in amniotic cells. In 1971 there were 3.56 million live births in the United States. Approximately 5,000 of these infants had Down syndrome, and 35 percent of those were born to women aged 35 or older. Thus, monitoring this group of pregnancies with amniocentesis would identify the syndrome prenatally in approximately 1,700 cases, but this would involve testing in nearly 250,000 pregnancies. The decision for or against amniocentesis and therapeutic abortion belongs to the patient; but a conservative estimate that at least half of the patients in 35-plus age group would request the procedure indicates the scope of prevention. Two problems emerge: the need to inform more physicians of the indications for amniocentesis, and our lack of facilities for handling such a potentially large number of requests for amniocentesis. Cost is another consideration, and the need for state and federal support must soon be faced. For those wishing it, at least, this expensive test should be free. A longer period of time must elapse before the risks (both immediate and future) of amniocentesis can be evaluated. However, it may eventually be possible to monitor all pregnancies for chromosome abnormalities in those requesting this procedure, so that Down syndrome will become a potentially preventable malformation syndrome.

Recent advances in the techniques of ultrasound scanning may soon provide a practical noninvasive technique for the detection of fetal anomalies by the 14th week of gestation.

An extension of amniocentesis is the use of radiopaque dyes to detect malformations such as clefts, spina bifida, and so forth. In the future, fetoscopy may be used in couples at high risk for a child with malformations. It may also prove safe enough for use in obtaining fetal blood samples, skin, and specialized tissues for studies not possible on amniotic cells, so that fetal surgery can be performed and metabolic disorders treated early enough to prevent irreversible effects.

The advances made in cytogenetic techniques in the past three years can be expected to continue, so that small, previously undetectable deletions and duplications will uncover further causes of mental retardation. Automated karyotyping, programmed to the position of chromosome bands or other markers, will improve the accuracy and efficiency of this technique. A chromosome analysis, employing larger cell counts than have hitherto been practical, will then become part of the work-up of every newborn.

Advances in biochemical techniques will result in the delineation of additional metabolic errors causing mental retardation. Although most of these, like the ones already described, are individually rare, screening tests for newborns will probably be implemented in most states. The wider use of carrier detection may allow the monitoring of the first pregnancy through amniocentesis, so that the fetus may be treated while still in utero. An increasing number of metabolic diseases can probably be corrected by using improved tissue culture techniques to study enzyme enhancement through cofactors and other substances. Other diseases can be prevented by means of enzyme replacement within a membrane (to avoid immunologic problems). Pregnancies in women who have been treated for metabolic disorders such as PKU, can pose problems. However, much more effective long-term surveillance and multiphasic screening tests early in pregnancy can probably prevent some of these cases of mental retardation.

Studies of the development of human hemoglobin have demonstrated that certain enzymes are probably active for a period of time in utero, only to be "switched off" and replaced by the definitive adult enzyme shortly before or after birth. It is known, for instance, that α-fetoprotein is produced by the liver and upper gastrointestinal tract of the fetus but disappears before birth. There are probably many other enzyme systems that follow a similar pattern of development. Mental retardation may, in some cases, be due to a defect in one specific enzyme which cannot be detected at birth. Biochemical studies of fetal proteins have rarely been attempted, but better understanding in this area could lead to prevention and replacement of the defective enzyme.

It is already becoming clear to many of us that genetic counseling must be made available to a far greater number of families. An increasingly well-informed public is beginning to make demands which will soon strain our limited resources. As long as most genetic services are offered only in large medical centers, they can only meet the needs of the few who know where to seek them. Satellite clinics are needed throughout the country, and they require planning at state and national levels. Although one of the basic requirements for genetic counseling, the diagnosis, can only be provided by the physician, the supply of physicians trained in genetics is insufficient to provide adequate counseling services and the necessary follow-up. Programs are needed to train nurses and others as auxiliary genetic counselors, particularly for purposes of follow-up. The ideal would be a team approach involving the physician, nurse, and social worker, with support from a psychiatrist and psychiatric social worker. In view of rising expenditures for the care of the mentally retarded, commitments are urgently needed for prevention through adequate genetic counseling, amniocentesis and screening programs, and additional laboratory facilities.

REFERENCES

1. Holmes, L. B., Moser, H. W., Halldorsson, S., et al.: *Mental Retardation: An Atlas of Diseases with Associated Physical Abnormalities.* New York: Macmillan, 1972.

2. Carter, C. H.: *Handbook of Mental Retardation Syndromes,* 2nd. ed. Springfield: Charles C. Thomas, 1970.

3. Penrose, L. J.: *The Biology of Mental Defect,* 2nd ed. New York: Grune & Stratton, 1963.

4. Slater, E. and Cowie, V.: *The Genetics of Mental Disorders.* London: Oxford University Press, 1971.

5. Carr, D. H.: Chromosome anomalies as a cause of spontaneous abortion. *Am. J. Obstet. Gynecol. 97:*283, 1967.

6. Walzer, S., Breau, G., and Gerald, P. S.: A chromosome survey of 2,400 normal newborn infants. *J. Pediatr. 74:*438, 1969.

7. Hall, B.: Mongolism in newborn infants. *Clin. Pediat.* (*Phila.*) *5:*4, 1966.

8. Darling, D. and Benda, C. E.: Mental growth curves in untreated institutionalized mongoloid patients. *Am. J. Ment. Defic. 56:*578–788, 1952.

9. Berg, J. M., Crome, L., and France, N. E.: Congenital cardiac malformations in mongolism. *Br. Heart J. 22:*331, 1960.

10. Richards, B. W., Stewart, A., Sylvester, P. E., and Jasiewicz, V.: Cytogenetic survey of 225 patients diagnosed clinically as mongols. *J. Ment. Defic. Res. 9:*245–259, 1965.

11. Edwards, R. G. and Fowler, R. E.: *Modern Trends in Human Genetics,* ed. A. E. H. Emery. New York: Appleton-Century-Crofts, 1970, vol. 1, pp. 181–213.

12. Fialkow, J.: Thyroid antibodies, Down's syndrome, and maternal age. *Nature 214:*1253, 1967.

13. Carter, C. O.: A life table for mongols with the causes of death. *J. Ment. Defic. Res. 2:*64–74, 1958.

14. Forssman, H. and Akesson, H. O.: Mortality in patients with Down's syndrome. *J. Ment. Defic. Res. 9:*146–149, 1965.

15. Oster, J.: *Mongolism.* Copenhagen: Danish Science Press, 1953, pp. 335–336.

16. Collmann, R. D. and Stoller, A.: Data on mongolism: prevalence and life expectation in Victoria, Australia. *J. Ment. Defic. Res. 7:*60–68, 1963.

17. Owens, D., Dawson, J. C., and Lesin, S.: Alzheimer's disease in Down's syndrome. *Am. J. Ment. Defic. 75:*606–612, 1971.

18. Carter, C. O. and Evans, K. A.: Risk of parents who have had one child with Down's syndrome (mongolism) having another child similarly affected. *Lancet 2:*785–787, 1961.

19. Stern, Z., Susser, M., and Guterman, A. V.: Screening program for prevention of Down's syndrome. *Lancet 1:*305, 1973.

20. Hamerton, J. L.: *Human Cytogenetics I.* New York: Academic Press, 1971, p. 279.

21. Edwards, J. H., Harnden, D. G., Cameron, A. H., et al: A new trisomic syndrome. *Lancet 1:*787, 1960.

22. Smith, D. W., Patau, K., Therman, E., and Inhorn, S. L.: A new autosomal trisomy syndrome. *J. Pediatr. 57:*338, 1960.

23. Crandall, B. F. and Ebbin, A.: Trisomy 18 and 21 in two siblings. *Clin. Genet. 4:*517–519, 1973.

24. Warkany, J., Passarge, E., and Smith, L. B.: Congenital malformations in autosomal trisomy syndromes. *Am. J. Dis. Child. 112:*502, 1966.

25. Caspersson, T., Linsten, J., Zech, L., et al.: Four patients with trisomy 8 identified by the fluorescent and Giemsa banding techniques. *J. Med. Genet. 9:*1–7, 1972.

26. Crandall, B. F., Bass, H. N., et al.: The trisomy 8 syndrome: two additional mosaic cases. *J. Med. Genet. 11:*393, 1974.

27. Crandall, B. F., Muller, H. M., and Bass, H. N.: Partial trisomy of chromosome no. 15 identified by trypsin-Giemsa banding. *Am. J. Ment. Defic. 77:*571–578, 1973.

28. Turner, H. H.: A syndrome of infantilism, congenital webbed neck, and cubitus valgus. *Endocrinology 23:*566, 1938.

29. Ford, C. E., Jones, K. W., Polani, P. E., et al.: A sex-chromosome anomaly in a case of gonadal dysgenesis (Turner's syndrome). *Lancet 1:*711, 1959.

30. Carr, D. H.: Chromosome anomalies as a cause of spontaneous abortion. *Am. J. Obstet. Gynecol. 97:*283, 1967.

31. Shaffer, J. W.: A specific cognitive deficit observed in gonadal aplasia (Turner's syndrome). *J. Clin. Psychol. 18:*403–406, 1962.

32. Hamerton, J. L.: *Human Cytogenetics II: Clinical Cytogenetics.* New York: Academic Press, 1971.

33. Race, R. R. and Sanger, R.: Xg and sex chromosome abnormalities. *Br. Med. Bull. 25:*99–103, 1969.

34. Ford, C. E., Jones, K. W., Miller, O. J., et al.: The chromosomes in a patient showing both mongolism and the Klinefelter syndrome. *Lancet 1:*709, 1959.

35. Knox, W. E.: Phenylketonuria. In Stanbury, J. B., et al. (eds.): *The Metabolic Basis of Inherited Disease.* New York: McGraw-Hill, 1973, pp. 266–295.

36. Perry, T. L., Hansen, S., Tischler, B., et al.: Unrecognized adult phenylketonuria. *N. Engl. J. Med. 289:*395–398, 1973.

37. Sloan, H. R. and Frederickson, D. S.: GM_2 Gangliosidoses: Tay-Sachs disease. In Stanbury, J. B., et al. (eds.): *The Metabolic Basis of Inherited Disease.* New York: McGraw-Hill, 1973, pp. 615–638.

38. Spranger, J. W. and Wiedemann, R.: The genetic mucolipidoses. *Humangenetik 9:*113, 1970.

39. Berg, J. M. and Crome, L.: Les phakomatoses dans la déficience mentale. In

Michaux, L. and Feld, M. (eds.): *Les Phakomatoses Cérébrales*. Paris: S.P.E.I., 1963, pp. 297–304.

40. Crowe, F. W., Schull, W. J., and Neel, J. V.: *Multiple Neurofibromatoses,* Springfield: Charles C. Thomas, 1952.

41. Fraser, F. C.: Genetics and congenital malformations. In Steinberg, A. G. (ed.): *Progress in Medical Genetics,* New York: Grune & Stratton, 1961, pp. 38–80.

42. Edwards, J. H., Norman, R. M., and Roberts, J. M.: Sex-linked hydrocephalus. *Arch, Dis. Child. 36:*481–485, 1961.

PART II

Genetic Counseling

7

PSYCHOLOGY AND METACOMMUNICATION IN GENETIC COUNSELING

Michael A. Sperber, M.D.

With innovation in genetic technology, and a changing moral and legal climate surrounding abortion and amniocentesis, human beings now have the capacity to modify their genetic makeup. Cross-disciplinary collaboration between psychiatry and genetics increases as genetic knowledge advances. Painstaking evaluation and planning are necessary to prevent any abuse of human rights resulting from misapplication of this rapidly expanding genetic knowledge.

Many areas of psychiatry and genetics converge in genetic counseling. While all interdigitate, several will be considered separately to facilitate discussion: the social control of biomedical (genetic) technology, the psychologic effects of this technology, eugenics, and the intrapsychic and interpersonal impact of a genetic defect.

Social Control of Biomedical (Genetic) Technology

The potential benefits of genetic research must be weighed against its potential hazards. More broadly, the issue is achieving the proper balance between freedom and control in a complex technologic society:

"The main fear associated with the area of human genetic research is that once sufficient knowledge is gained, techniques will become available that will allow men to manipulate other men. The assumption is that people will be deliberately misled or coerced into allowing themselves to be manipulated. In part, there is the concern that some degree of voluntary submission to manipulation might lead ultimately to involuntary controls imposed in the interest of those who govern." [1]

There is a need for social controls that will safeguard traditional human rights without stifling research in genetics. "In medicine we are called upon to define those personal values that society will not relinquish for any material good and those which can be invaded for a worthy social purpose." [1] Psychiatry may play a role in identifying and defining inviolable values, and the genetic counselor is in a position to influence considerably the client's value system and to affect decision-making processes.

Psychologic Effects of Genetic Technology

It is too soon to determine the psychologic effects of genetic technology, but research in this area is necessary. For example, in states requiring preschool sickle-cell identification, at what stage would it be preferable, from a psychiatric standpoint, to relay a positive diagnosis—in early childhood or in adolescence, before or after marriage? As Lappé writes,

> the simple act of acquiring prenatal genetic information about a fetus . . . automatically sets into motion a train of events which themselves change that individual's future. . . . You tell him (and sometimes others) something about where he came from and who is responsible for what he is now, you project who he may or may not become in the future. You set certain limits on his potential. You say something about what his children will be like, and whether or not he will be encouraged or discouraged to think of himself as a parent. In this way the information you obtain changes both the individual who possesses it, and in turn the future of that information itself.[2]

Should the prospective marriage partner be informed? Could doing so affect an impending marriage or an ongoing one? Other areas for psychiatric research include the effect on a family of using an artificial insemination donor, and the influence of a particular psychologic organization on an individual's or a family's ability to tolerate the stresses of genetic defects in general, or of a particular type of genetic defect.

Eugenics

Positive eugenics is the preferential breeding of "superior" individuals in the population to maintain or improve genetic stock by means of, for example, a sperm bank using genetic studs with "desirable" intellectual and social traits and without physical handicaps.[3] Negative eugenics involves discouraging or legally prohibiting reproduction by individuals carrying genes leading to disease or disability.

When applied to populations, the use of positive and negative eugenics necessitates determining what is genetically "good" or "bad." Who will judge, and how will they distinguish "good" genes from "bad" ones? The pitfalls of positive eugenics were demonstrated in Nazi Germany. It is important to realize that what is "good" at one time may not be in an altered future environment. Well-meaning geneticists, seeking to change the gene pool, might choose to delete a desirable heterogeneity. Also, groups of genes may have been selected together (co-adapted), and if one is excised, the effects, desirable or undesirable, positive or negative, may be greater than desired for the whole group of genes.

Who shall determine what the future species will be like? What characteristics of intellect and behavior should be bred into future generations, and which bred out? Who shall be allowed to procreate or to have the benefits of transplanted or artificial organs? The psychiatrist, whom society considers to be best prepared to deal with the total human being, may well be called upon to establish these intellectual and behavioral norms, an incredibly complex task.

Psychiatry and Genetic Counseling

The functions of the genetic counselor include: (1) diagnosing and describing pathogeneses and complications of a given disease; (2) calculating the odds and probabilities that a genetic defect will occur; (3) helping the client reach a decision in the light of an acknowledged genetic risk and within the religious, social, ethical, legal, economic, and cultural contexts of the situation; and (4) describing treatments and community resources.

Genetic counseling may drastically change lives and relationships. Therefore, as Rainer states, "it may very well have to be the job of some psychiatrists to be caretakers of the human aspects of genetics, to have a voice in the larger issues and, if they do not do all the counseling themselves, at least to be concerned and to supervise and to be involved with it." [4]

Several areas of special significance to psychiatrists are not susceptible to facile solutions. For example, there is disagreement about whether the counselor should use a normative approach or should state the risk estimate as neutrally as possible, without attempting to influence the counselee's decision.*

In addition, communication often breaks down. For example, Opitz observes that "when some parents are first told that their child has a severe genetic disease they often exhibit hostility and repression. Their recall is poor. In short, they are simply not ready or willing to accept what I tell them." [5] Leonard's survey reports various communication problems with nearly 50 percent of 61 couples counseled.[6] Five families indicated that they had never been counseled. Six others did not accept certain parts of the counseling. Five could not remember what was said. Nine other families could not understand the counseling or apply it in a meaningful way to their situations. Thus, counselees appear to employ mechanisms of denial, repression, and reaction formation when dealing with the highly charged emotional material conveyed in the counseling situation.

* Opitz states: "I avoid any 'selling' like the plague. It's never possible to put oneself entirely into anyone else's shoes, and I feel that the integrity of the couple's decision-making must be vouchsafed." [5]

A complex issue is that of confidentiality.* In particular, mass genetic screening may lead to leakage of confidential data, owing to the loss of strict privacy inherent in more closely circumscribed doctor-patient relations. What should the counselor do when he learns that a member of his client's family may be at risk? Should the marriage partner or a relative be informed that an individual has Huntington chorea? Who should be told, and when, if a neonate is found to have an XYY chromosome complement? Gardner notes: "Inappropriate discussions by counselors with parents may interfere with [the infant's] normal gender role development . . . and may, indeed, produce pathological changes in [him]. The counselor must be aware that he uses himself as a therapeutic instrument in terms of his relationship with parents and infants." [7]

Transmitting certain genetic information may well create great stress in families.† For example, in studies of hemophiliacs, Agle [8] notes maternal guilt and overprotection, as well as overinvolvement of family resources, often to the detriment of other family activities.

When one partner is involved in transmitting a genetic defect, the genetic counselor must decide whether or not to impart this information. In the opinion of Stevenson et al.,[9] "it seems wise, unless it is obvious to patients or necessary for some reason, not to indicate that a condition was inherited through one or the other parent, as this can be a source of marital friction." (This view might be considered infantilizing by others.) Carter's data show that disclosing carrier status causes changes in a couple's reproductive plans, as well as a change of partners in some instances.[10] In Leonard's study of 61 couples with genetic problems (cystic fibrosis, phenylketonuria, Down syndrome), 23 (about 38 percent) reported that genetic counseling had had an adverse effect on their sex lives.[6]

Birenbaum notes that parents of the mentally retarded are in an ambiguous position: although their reproductive capacity shows no evi-

* See also Chapter 10, pp. 241–242.

† Sorenson's essays provide scholarly research and good references. See J. R. Sorenson: *Social Aspects of Applied Human Genetics*. Russell Sage Foundation, 1971. See also "Genetic Counseling: Some Psychological Considerations." Paper read at AAAS, Washington, D.C., December 29, 1972. Certain points raised in this section derive from Sorenson's discussions.

dence of biologic abnormality, they consider themselves abnormal and are considered so by society.[11]

Parents may come to perceive their family as "doomed," an attitude that can lead to depression and disorganization of the family.[12] Tips describes a family's reaction after being informed of a genetic defect: ". . . The parents [became] concerned about their interpersonal relationships. Nocturnal arguments had ensued and the mother had developed severe dyspareunia. She had become overly concerned with the management of the affected children and had neglected the older children, who developed behavioral disturbances and began failing in school. The father, previously dominant, had withdrawn and assumed a passive role in the family circle." In another case, "the patient's family had been apparently well and adjusted prior to his diagnosis. . . . Quarrels between parents concerning interpersonal psychosexual behavior and genetic prognosis had ensued almost immediately thereafter. The family physician and the parents had become aware that the 'intellectual' debates were causing tension and anxieties among all the offspring." [13] Other problems may arise from amniocentesis, since abortion is not universally accepted. Miller asks, "What are we going to do in situations where the husband and wife don't agree on this issue, and one wants their abnormal fetus to be aborted while the other does not?" [14]

Feelings of low self-esteem and defectiveness are often major issues confronting those afflicted with genotypic or phenotypic disease.[15] Lax observes that "the procreation of an impaired child profoundly affects the mother's self-image, causing a severe decrease in the magnitude of positive self-directed feelings, resulting in a profound feeling of worthlessness. . . ." The mother reacts with "disappointment, narcissistic mortification and depression. . . ." Her depression, revealed by the extent of her conscious or unconscious rejection of the child, affects her maternal behavior and the child's developing self-image. Patterns of mother-defective child interaction "encompass the entire gamut from oversolicitude to complete rejection; from apparent blindness and insensitivity to the child's handicap to an exaggerated over-magnification of the defect . . . from granting the child the prerogatives of an exception to treating the child as the scum of the earth." [16] Faced with the possibility of a psychic breakdown, the mother may deny the existence of the defect. "Rejection of various degrees, culminating in the acutal giving away of the child, is a reflection of the mother's need to emphasize her separateness from the child who symbolically represents her unconscious defective self." [16]

Forrer describes a mother's obsessional, frantic devotion to her mentally and physically defective child, as a consequence of which her other, healthy children and the family unit are neglected or disregarded.[17]

These observations have been borne out by recent studies.[18] "Anxiety, hostility, and depression levels were significantly higher in parents seeking counseling than in normative controls (both $P \leq .002$). Following genetic counseling, there was significant improvement in anxiety ($P < .0005$) and depression ($P < .05$) along with significant improvement in overall self-concept ($p < .01$)." [18]

The physically handicapped and their families may experience guilt and shame in varying degrees. Guilt, which arises when the superego is in conflict with the ego, may be the initial response to the birth of a defective child. The parents may ask what social, religious, legal, or moral sanction was violated that would account for such a calamity. Parents often handle guilt by projecting culpability. Blaming others—their doctors, their spouses—alleviates their guilt. Some parents scapegoat their handicapped offspring, compounding the latter's difficulties in adjusting.

Shame, which results from a discrepancy between ego and ego ideal, is more usually experienced by the afflicted than by their parents. In our society, where much emphasis is placed upon unblemished beauty, the individual experiences shame when he feels he falls short of familial norms or social ideals. Shame, accompanied by increased heart rate, blushing, and sweating, is difficult to defend against. The individual may feel mortified and withdraw from his fellows.

W. Somerset Maugham [19] sensitively describes these responses in his thinly veiled autobiography, *Of Human Bondage*. Philip Carey, the hero, is afflicted with talipes (Maugham had an intractable stutter), of which

he had grown horribly sensitive. He never ran if he could help it, because he knew it made his limp more conspicuous, and he adopted a peculiar walk. He stood still as much as he could, with his club-foot behind the other, so that it should not attract notice, and he was constantly on the lookout for any reference to it. Because he could not join in the games which other boys played, their life remained strange to him . . . and it seemed to him that there was a barrier between them and him. . . . He was left a good deal to himself. . . .[He] passed from the innocence of childhood to bitter consciousness of himself by the ridicule which his club-foot had excited.

Not all persons with a congenital handicap experience shame because of it, however. Thus, as a medical student, Philip studies a young patient with the same pedal malformation as his own. "Philip examined the foot and passed his hand over the shapelessness of it. He could not understand why the boy felt none of the humiliation which always oppressed himself. He wondered why he could not take his deformity with that philosophical indifference.

The identity of a person who perceives himself as psychologically defective may crystalize around his physical handicap, which to him symbolizes his psychologic deformity. In attempting to reduce the stigma of an impairment, genetic counselors and therapists would do well to enhance the handicapped person's feelings of competence and sense of self-worth. As Philip develops more assurance about his value to others, he becomes better able to take his defect in stride:

He accepted the deformity which had made life so hard for him; he knew that it had warped his character, but now he saw also that by reason of it he had acquired that power of introspection which had given him so much delight. Without it he would never have had his keen appreciation of beauty, his passion for art and literature, and his interest in the varied spectacle of life. The ridicule and the contempt which had so often been heaped upon him had turned his mind inward and called forth those flowers which he felt would never lose their fragrance. Then he saw that the normal was the rarest thing in the world. Everyone had some defect, of body or of mind. . . . He thought of all the people he had known, (the whole world was like a sickhouse, and there was no rhyme or reason for it.) He saw a long procession, deformed in body and warped in mind, some with illness of the flesh, weak hearts or lungs, and some with illness of the spirit, languor of will, or a craving for liquor. At this moment, he could feel a compassion for them all.[19]

Having overcome his shame, Philip is able to move out of the masochistic bondage characterizing his relationship to one woman and form a loving, mutually satisfying relationship with another.

In working with the congenitally handicapped, the genetic counselor must be sensitive to interpersonal as well as intrapsychic issues. If he comes to grips with his own feelings of defectiveness and his biases about deformity in general, he will find it easier to help his patients master guilt and shame and assume the task of autonomous decision making.

REFERENCES

1. Mendelsohn, E., Swazey, J. P., and Taviss, I. (eds.): *Human Aspects of Biomedical Innovation.* Cambridge: Harvard University Press, 1971, pp. 33, 26, 38.

2. Lappé, M.: Moral obligations and the fallacies of "genetic control." *Theological Studies 33:*412, 1972.

3. Muller, H. J.: Guidance of human evolution. *Perspect. Biol. Med. 1:*590, 1959.

4. Rainer, J. D.: Genetic counseling, social planning and mental health. In Redlich, F. C. (ed.): *Social Psychiatry: Proceedings of the Association.* Baltimore: Williams and Wilkins, 1969, p. 229.

5. Opitz, J. M.: Genetic counseling: is it heeded too little or too much? *World Med. News,* Sept. 8, 1972, pp. 17, 18.

6. Leonard, C., Chase, G., and Child, B.: Genetic counseling: a consumer's view. *N. Engl. J. Med. 287:*433, 1972.

7. Gardner, L.: Counseling in genetics. In Harris, M. (ed.): *Early Diagnosis of Human Genetic Defect: Scientific and Ethical Considerations.* Washington, DC: HEW Publication No. (NIH) 72–25, 1972, p. 191.

8. Agle, D. P.: Psychiatric studies of patients with hemophilia and related states. *Arch. Intern. Med. 114:*76, 1964.

9. Stevenson, A. C., Davison, B., and Oakes, M.: *Genetic Counseling.* Philadelphia: J. B. Lippincott, 1970.

10. Carter, C. O., Evans, K., Fraser-Roberts, J., and Buck, A.: Genetic counseling: a follow-up. *Lancet 1:*281–285, 1971.

11. Birenbaum, A.: On Managing a courtesy stigma. *J. Health Soc. Behav. 11:*196, 1970.

12. Langsley, D. G.: Psychology of a doomed family. *Am. J. Psychother. 15:*531, 1961.

13. Tips, R. L., Smith, G. S., Lynch, H. T., and McNutt, C. W.: The "whole family" concept in clinical genetics. *Am. J. Dis. Child. 107:*67, 1964.

14. Miller, O. J.: An overview of problems arising from amniocentesis. In Harris, M. (ed.), *Early Diagnosis,* p. 23.

15. Geis, H.: The problem of personal worth in the physically disabled patient. *Rehabil. Lit. 33:*34, 1972.

16. Lax, R. F.: Some aspects of the interaction between mother and impaired child: mother's narcissistic trauma. *Int. J. Psychoanal. 53:*339–341, 1972.

17. Forrer, G. R.: The mother of a defective child. *Psychoanal. Q. 28:*59–63, 1959.

18. Antley, R. M. and Hartlege, L. C.: Psychological responses to genetic counseling for Down's syndrome. (In press.) See also Antley, M. A., Antley, R. M., and Hartlege, L. C.: Effects of genetic counseling on parental self-concepts. *J. Psychol. 83:*335, 1973.

19. Maugham, W. S.: *Of Human Bondage*. New York: Vintage, 1956, pp. 44–48, 460, 680.

8

GENETIC COUNSELING: A PSYCHOTHERAPEUTIC APPROACH TO AUTONOMY IN DECISION MAKING

Marc Lappé, Ph.D. and
Julie Anne Brody, M.A.

Counseling and the Control of Behavior

The first axiom of behavior control * is that control is a consequence of power. Because every professional is a bridge to an otherwise inaccessible body of knowledge, there will always be a power differential in the kind of professional-nonprofessional relationship that exists in genetic counseling. Whether or not genetic counseling ultimately restricts or enhances the counselee's decision-making autonomy will depend upon (1) the extent of the counselor's control, (2) the counselor's (and coun-

* The pejorative sense of "behavior control"—coercion, force, threats of punishment, deception, or other modes of misrepresentation—need not be operative for an activity to have a high potential for control. For example, we would construe "behavior control" to embrace an interaction which reduces or restricts the action options available to one or both parties. We would similarly regard an interaction which *forces* actions which might have remained unresolved to be a form of behavior control.

selee's) consciousness of the potential for controlling, and (3) the value matrix of the help-giving professional.

In an absolute sense, some exogenous control over counselee behavior is inevitable in genetic counseling because the counselor has potentially total control over incoming stimuli. On a relative scale, it is neither inevitable nor necessarily undesirable. There would be no rationale for genetic counseling if it were not to some significant extent to "influence" people's behavior. Such influence may be directed solely at changing the basis for decision-making behavior of a couple (from intuitive to probabilistic), or directed toward changing the reproductive behavior of the couple itself. In either instance, it is necessary to specify at what point enlightened influence becomes nefarious control. The distinction between influence and control is a matter of degree; perhaps the limit of influence on a conscious level is the point at which it comes overtly or covertly to prescribing actions for the counselee. The distinction between enlightened and nefarious influence perhaps comes at the point where the individual's rational processes (and hence autonomy) are superseded. This criterion has been used by at least one psychiatrist to define behavioral influence which is coercive: "If you exploit the anxieties of an individual rather than his reason, and this may be done with his awareness or without his awareness, you are coercing behavior, often without either the recognition or admission of coercion on either part." [1]

We would maintain that the distinction between a desirable influence (to be specified below) and undesirable control must include unconscious factors which might contribute to directed actions for the counselee. *The danger of undesirable (i.e., coercive) "behavior control" in the context of genetic counseling lies not in the threat of maliciousness or deliberate manipulation, but in the ignorance of how controls can be unintentionally exerted.* For example, genetic counseling will often affect the counselee's decision making by modifying his perception of his problem. His own perception may be reinforced or disrupted by what he consciously believes he perceives as the counselor's "motive," as well as a constellation of surd, irrational, and unconscious elements in the counseling process. We must be concerned with the total experience of counseling as it affects the counselee, as well as the "control"-laden events of information giving and receiving. We will therefore deal with at least four categories of variables in counseling: (1) interaction or process variables, (2) physical and psychosocial milieu variables, (3) counselor variables, and (4) counselee variables.

Interaction Components of Counseling: Biasing Factors in the Content and Method of Genetic Counseling

The crux of genetic counseling lies in the presentation of relevant genetic information. An information giver always has the potential for controlling the behavior of the information receiver because he can regulate stimulus input. Specific dangers associated with the methodologies of information giving in genetic counseling transcend the obvious pitfalls of misdiagnosis, ambiguous diagnosis, and plain misunderstanding. While the status of our genetic knowledge and the nonspecific nature of many genetic diagnoses suggest a high potential for factual misunderstanding, we believe that the mental sets of the counselee and counselor also determine the effectiveness of information exchange.

Ineffectual or misleading communication can occur in genetic counseling when information is introduced at a time in the counseling process when the counselee is psychologically unable to assimilate it (for example, because of anxiety or guilt). It may also occur when information is presented in a form or language which is incomprehensible to a counselee because he has had little experience with genetic disease, or because his cultural experience has not taught him to think in statistical terms.

A second level of behavioral influence can be linked to the quantity of total information given to the counselee. A genetic counselor may overwhelm the counselee with large volumes of data ("information overload"), or compromise the counselee's control over the situation by withholding information. For example, a counselor may choose to conceal, in an effort to offset the "damaging" impact of the information, which one of the parents of an affected child is carrying the responsible gene, or to withhold the results of a chromosomal analysis which indicates the presence of an ambiguous condition in the fetus.[2,3] While these decisions can be rationalized, the precedent of nondisclosure offers grounds for more severe kinds of control by selective information limitation.

We believe that the core potential for controlling behavior (for good or for ill) in genetic counseling resides in the process of counseling itself. This process includes the counselor's method of conveying information to the counselee, his own goals in counseling, and how he works with the counselee to achieve those goals. A simplified description of the counseling process depicts three general approaches:

(1) Action-oriented or directive counseling, in which the counselor participates in the decision-making process and offers advice in varying degrees toward the resolution of the counselee's problem.
(2) Insight-oriented or nondirective counseling, in which the goal is client self-determination and the counselor refrains from directly entering the decision-making process. Directive advice as such is not offered. The counselor's aim in this situation is to aid his client in making a choice rather than to indicate which choice is appropriate.
(3) Information-giving counseling, in which the counselor views himself as value-neutral. The counselor in this guise will never consciously enter the decision-making process and usually perceives his goal solely in terms of giving "accurate" genetic information to the counselee. In such circumstances, the counselor may perceive himself as a genetic educator rather than as a facilitator or an advice giver.

Undesirable behavior control is possible in any of these models. "Control" may be more overt in any directive approach to counseling; yet, paradoxically, overt control may be less coercive than "nondirective" counseling. In the latter case, the counselor's motives may be less explicit. An "educational" approach may also have a high potential for control by virtue of its stated "neutrality." In disregarding the steps which are necessary to enable the client to assimilate the information to his own life-style, relationships, values, goals, and needs, an information giver may be coercive because he avoids the resolution of (or may even create) emotional obstacles to understanding genetic information. The consequences of the counselor's inability or deliberate refusal to deal with the highly charged emotional issues which surround genetic disease or carrier status may include a reduction in the counselee's self-esteem, anger toward the counselor, or even denial of the validity of the genetic information itself.

Biasing Factors in the Set and Setting of Genetic Counseling

PHYSICAL MILIEU AND HOSPITAL SUBCULTURE

Genetic counseling is commonly understood as an instance of medical advice giving. Moreover, genetic counseling units have increasingly come to be located in a medical or hospital setting. Such an institutional atmosphere implies that something is wrong, that there is a corrective procedure to be followed, and that compliance is expected. The physical milieu creates an initial potential for behavior control because the medical setting is associated with the doctor-patient mode of communication, in which the doctor's traditional role is an authoritarian, prescribing one. Beyond this, the medical context connotes illness and deviance—a place where patients usually come for treatment and cure.

Whether the genetic counselee is referred or self-referred will undoubtedly make a large difference in his perception of the counseling experience; whether counseling is with a couple or a single parent is also important. Beyond this, the context in which genetic counseling is offered contributes to the client's perception of the counselor as prestigious, credible, and powerful. The setting may also lead to the counselee's expectation of direction and advice. To the extent that the genetic counselee assumes the patient role, he may be limiting his normal decision-making responsibility. In this sense he may abdicate some of his autonomy, assigning responsibility for a series of ego functions to the counselor and perhaps to the institution itself. Thus, his freedom to refuse genetic advice, to act in terms of his own values and interests, and to question the procedures to which he will be subjected is diminished. A medical milieu may reinforce compliant, dependent, and passive behavior.

Counseling itself may take place in a physician's office or in a multipurpose room within a hospital. At the outset of counseling, a pedigree and family medical history are taken. The counselor, physician or not, often distinguishes himself from the counselee by means of a white lab coat and clipboard. An examining room and facilities for diagnostic work-up are an integral part of this milieu. This microenvironment, which resembles that in which the sick are ministered to, may reinforce the client's tendency to feel dependent as well as deviant. These feelings may be accentuated by stereotypic nonverbal cues, such as seating arrangement, body posture, telephone interruptions, and the counselor's tone of voice, eye contact, gestures, and facial expressions. The short time allotted to counseling (i.e., both the length and number of sessions, as much the result of cost and scheduling as of design) interferes with the client's appropriate assimilation of the counseling information and emphasizes the social power gulf between anxious and/or concerned client and busy counselor.

DIFFERING CHARACTERISTICS OF COUNSELORS AND COUNSELEES

Differences in social and cultural background, including ethnic and religious experience, also contribute to the gap in counselor-client communication. Major differences in values and value systems have been demonstrated in at least one doctor-patient model, that of the seeker of contraceptive advice and her physician.[4] The related pseudopatient status of many counselees (their pregnant or carrier roles), their lack of ability to quantitate risks, and their often desperate dependency may foster an obligation (perhaps unconscious) to follow what they perceive to be the counselor's "genetic advice" while disregarding their own values. Alternatively, they may defensively reject the perceived advice.

The effects of such value conflicts may be intensified when the counseling concerns an ethnically related disease foreign to the counselor. Some of the screening programs for sickle-cell anemia currently illustrate this kind of conflict. Indeed, some black people regard genetic counseling (and contraceptive education) in the milieu of sickle cell anemia as potentially genocidal. Specific demographic variables and socially determined prejudices can influence the counselor's behavior as well. For example, he may be unaware of the role which societal, staff, or hospital administration pressures play in his own interaction with the counselee, or even that his understanding of the problem differs from that of his client. Thus, genetic counseling also has a potential for discriminative control of reproductive freedoms predicated on a racial or ethnic basis.

The ways in which sociologic or cultural variables affect the counseling interaction, and thus the likelihood of behavior control, are frequently unconscious. The psychodynamic configurations of counselors and client, their past histories, unconscious feelings about deformity or defying "God's will," or apprehension about counseling blacks (or whites) will influence the ways in which counseling will modify behavior in practice.

In a medical setting, clients may tend to behave in ways which will elicit approval from the counselor. The extent to which the counselor reinforces a client's need for dependency will determine in part what the client does with the genetic knowledge he gleans from the counseling experience.

The combination of high distress level and a sense of the counselor as "omnipotent" place the counselee in a particularly vulnerable position. He may thus be psychologically unprepared for the kind of information he is to receive, and he may incorporate it into his defense system in a manner totally incongruent with the goals of the counselor. For example, his repressed anger at being unable to cope with problems beyond his ken may manifest itself in undisguised aggression toward the counselor ("acting out"). Such reactions to the counseling process itself may block the counselor's ability to deal with the seemingly "irrational" reactions to the genetic information which he conveys—such as a client's occasional insistence on reproducing in the face of high risk. The possibility exists that the client may use the counselor's information in other ways—for example, as a way out of neurotic difficulties associated with reproduction. There are enough reported instances of counselees ending a marriage, terminating a "safe" pregnancy, or electing sterilization in the aftermath of counseling to suggest that such events may frequently occur.

Characteristics of the Counselee

COUNSELING AND THE "SICK ROLE" IN SOCIETY

The genetic counseling encounter is different from the traditional encounters of doctor and patient in the medical model. In the latter instance, people who define themselves as sick (or are so defined by others) seek help from a professional who is trained to serve as a bridge to some body of information necessary for restoring health. The person labeled as sick or deviant thus enters a therapeutic relationship with a person who is designated by his social role as a help giver. In response, the help seeker assumes the role of patient or client, thereby reducing his normal social obligations and responsibility for self. Beyond this, the patient role per se diminishes one's social power, both subjectively and objectively. Patients' lack of medical knowledge, low socioeconomic status relative to physicians', and low social status within the clinic may combine to define further their powerlessness vis-à-vis the counselor.

What occurs during the course of the doctor-patient relationship is often outlined in terms of the medical model of illness and treatment. A diagnosis is made, causes are sought, the agents of illness are identified, and therapy is prescribed. The severity of the illness, the restrictiveness or length of treatment, the prognosis for the future of the individual and his or her family, and especially the manner in which these facts are communicated to the patient (i.e., paternalistically, factually, sympathetically, etc.) will influence (if not by direct order, then by hidden value assumptions or prejudices) the subsequent behavior of the help seeker.

While all of this is to an extent part of genetic counseling, the nature of genetic disorders does not precisely fit the medical model. In genetic counseling, the counselor and the counselee do not usually contract to engage in a therapeutic exercise. A treatment or "cure" for a particular genetic dilemma may not exist—short of the decision to bear or not to bear one's own child. Thus, the decision itself becomes the therapeutic object of genetic counseling; and a decision not to reproduce is more "therapeutic," if one focuses solely on the genetics of the problem, than is the decision to "take the chance." It may be less "therapeutic" if the focus is on the totality of the client's emotional life. Where selective abortion is possible and the counselor's allegiance is to the prevention of genetic disease, there is a high risk of directive, behavior-controlling interactions in the counseling process.

Paradoxically, the counselee himself may feel in need of behavior control. The person who seeks his first contact with the counseling professional may be in a state similar to the anomie described by Durkheim.[5] As an individual confronted by an initially insoluble problem, he may not recognize any of the traditional problem-solving guideposts or supports offered by family, society, or friends. His dilemma is compounded by the breakdown of the extended family, which previously afforded much of the needed support in such situations. In this sense, then, genetic counseling requires collaboration with the counselee in creating new behavioral guidelines and value structures for identifying appropriate attitudes or courses of action for solving the particular kind of problem that besets him.

DEFINING THE SPECTRUM OF MIND-SETS THAT TYPIFY PROSPECTIVE COUNSELEES

In 1971, fully 87 percent of those seeking counseling had already experienced the birth of a child with a condition having a suspected genetic etiology. Eighty percent of the total had been referred through the agencies of a general physician, obstetrician, or other medically oriented person.[6] One might therefore anticipate that the expectations of a prospective counselee would be predetermined to some degree by these powerful external factors.

Ample literature describes the psychologic discontinuities and traumas accompanying the birth of a severely handicapped or "defective" child, which is often experienced as a grievous loss.[7] Instead of the healthy child that all (but especially the mother) had expected, there is an unwelcome, unanticipated, and frequently unaccepted intruder. Parents in such circumstances may come to the genetic counseling setting in dire need of therapy directed toward restructuring their perspectives about their own normalcy and their capacity to bear "normal" children.

The problem of the psychiatrist qua counselor in such a setting has been described as that of providing a milieu in which mourning can occur.[7] A genetic counselor in a similar setting might do well to act similarly toward the counselee, who perceives the birth of a defective child as a loss. The danger of failing to recognize such a syndrome lies in ignoring a potentially major psychologic impediment to adequately perceived counseling. The counselor who "has not sensed or understood the need that parents have to grieve about their tragic 'loss,' . . . will feel ineffectual and reproached by the parents when [the parents] indicate their need for repeated opportunities to review and to re-examine the past in the current 'loss.' "[7] Such parents may have feelings of loss, defeat, and resentment which distort and preset their expectations about the counseling interview. These feelings disturb the clear communication that must accompany effective genetic counseling.

The birth of a defective child may be a uniquely traumatic event for the mother, profoundly affecting her self-image. As one psychologist has observed, "Silently or aloud, the mother of such a child asks: 'What is wrong with me that I gave birth to such a child? Why has it happened to me? What have I done?'[8] Further, there is documentation of the extent to which her need to reject her defective offspring erodes normal maternal behavior. (See Chapter 7.)

The genetic counselor may also expect to be confronted by a subset of clients that includes parents who are deeply troubled by an inability to cope with events that have undermined their sense of control over their own lives. For some of these parents, the birth of a defective child may have been their supreme undoing. At least one psychoanalyst has described a "doomed-family syndrome" in which the members feel utterly at the mercy of fate, doomed to have one defective child after another.[9]

While not all counselees come to the counselor with fatalistic feelings, genetic counselors who probe the initial expectations of prospective counselees may discover that most have little sense of the causal events which underlie their dilemma. Some parents will continue to ask for attributions of causation which will allow one or both parents to accept the burden of guilt, in the face of the counselor's description of the events which actually underly their problem.

Characteristics of the Counselor

The counselor's professional identity will play a pivotal role in determining the manner he routinely adopts in counseling. Counselors may come from a variety of disciplines, such as medicine, psychology, genetics, theology, and social work. Their primary professional roles may include those of clinician, investigator or educator, psychotherapist, or often some combination of these. Each occupational identity and professional role carries with it some implicit approach to genetic counseling rooted in a value construct. Value orientations among counselors may include attributing high valences to societal good, reducing genetic disease, improving the gene pool, maximizing human potential, or maintaining the integrity of the family and the individual. (One of us [M. L.] has dealt at length with a construction of the set of allegiances that typify human geneticists and counselors.[10]) These allegiances, while largely unconscious, will influence the counselor's perception of his client's behavior and those aspects of it which he believes need modification or reinforcement.

To the extent that there is no well-established role definition for the genetic counselor, the potential for coercive behavior control in genetic counseling remains high. Without such a definition or, at a minimum, the counselor's self-declaration of value orientation, prospective clients will be unable even to match their value sets with those of their counselor, much less anticipate the ways in which their needs and expectations coincide with the norms of genetic counselors. The absence of a common set of norms or values, of an established "ethic" of intervention in genetic counseling, and especially of role constraints—these yield a high potential for behavior control. Who should counsel? How should counselors be trained? What is the primary goal of counseling? All these questions are now open to individual interpretation. The consequent ethical confusion about counseling per se reflects an inherent instability in the field, which may drive counselors towards "safer" modalities of counseling, emphasizing nondirectedness or information giving rather than a more overtly directive approach.

It is possible to find many genetic counselors who assert the value-neutrality of their counseling. Because they assiduously avoid breaching the domain of the decision-making process, they believe their counseling skirts the dangerous periphery of behavior control. Sheldon Reed, the progenitor of the genetic counseling movement in the United States as we now know it, was perhaps the first advocate of noninterventionism in genetic counseling. He believed that providing simple factual information was the basic (and sole) acceptable function of the genetic counselor. To go further, he maintained, was to go beyond what was required.[11]

Ten years of counseling experience have shown this position to be largely untenable in practice. Value-neutral genetic counseling is as unobtainable in the practical setting of the counselor's office as is value-neutral physics or chemistry. At some point, values apply, even if merely to justify expending efforts in this area rather than another.

Some genetic counselors accept this premise and consciously adopt a directive role. Fraser is one of many who now feel that the counselor must engage the counselee in the decision-making process "to help the counselee reach a wise decision in the light of the many confusing and often conflicting factors which may enter into the situation." [12] Another experienced counselor states: "I never tell parents what to do but I am prepared to say what I would do if I was in their place. Where the risk is only a moderate one I often volunteer the information that 'in their place I would not take the risk too seriously.' " [13]

Yet the question of directive vs nondirective counseling, or neutrality vs biased interventionism, remains a knotty one. A medical ethicist has summed up the current situation in the following way:

> It appears that there currently exists an impasse, a dilemma in role delineation. The "value neutral" role model is impossible both in practice and theory. Functioning as if value neutrality were possible might create a situation where needed checks on the counselling process are ignored. Furthermore, to make the counsellor into a data transmitting technician removes him from the responsibility of being a moral man. On the other hand, the "advice-giving" model necessarily tends to impose the professional's values on the patient who is thus deprived of responsibility as an autonomous moral agent.[14]

Genetic Counsel as a Form of Generic Counseling

We would like to suggest a putative model for genetic counseling which may help to resolve this impasse. In doing so, we will stress the need for further incorporation of a psychodynamic viewpoint into genetic counseling.

Until now, definitions of genetic counseling have usually emphasized the "genetic" component of the counseling interaction over the "counseling" aspect. A sociologist who has exhaustively studied the process and content of counseling describes it as "the presentation of information to an individual or a couple about their genetic or chromosomal health or the probable or possible genetic or chomosomal health of their progeny." [6] Others, including psychiatrists, have included interactional components in their definitions, but all too often the object of counseling remains genes, not the psyche.

Genetic counselors who perform a statistical calculus of "recurrence risks" and describe risks in probabilistic terms to counselees may totally obfuscate the necessary facts. We would suggest that one primal reason for the incomprehensibility of recurrence risks is rooted in existential questions. The "recurrence" of the birth of a defective child may be initially perceived as one which only fate controls. In the case of particularly serious birth defects, to perceive otherwise may be tantamount to accepting personal responsibility for something which is among the most incomprehensible of human events. When the counselee is asked to join with the counselor in calculating "risks" for such happenings, it may only be the counselor who can know that these events are merely statistical, nonteleogical happenings.

Like most people, the counselee may have a deep psychic need to attribute meaning to events. If he is forced to think about "recurrence risks" before expunging the normal psychic resistance to a probabilistic universe, his cognitive processes may be blocked. Some individuals may be locked into thinking solely in binary terms ("either this will happen again, or it won't") and may thus be condemned to continue to seek a (usually imaginary) chain of events which caused the original defect.

Fortunately, the advent of prenatal diagnosis through amniocentesis affords a means for bringing certainty to some counseling situations. Even with amniocentesis, however, the psychologic problems of the expectant parents are often not mitigated. A theologian has presented data which demonstrate the extent to which the decision to terminate a life in utero brings prospective parents to the brink of what it means to be human.[15] Thus, an existential dimension remains even here.

This type of background information provides some insight into the paradoxically poor results of many counseling interactions. For example, in a recent study describing the results of counseling for congenital heart disease, 75 percent of the 35 couples sampled could not remember the recurrence risk that they had been given one to four months previously.[16]

Recommendations

What then is the way out? We believe that genetic counseling can ensure the counselee's maximal autonomy by providing an atmosphere in which counselor and counselee can collaboratively articulate the covert questions, the anxieties, guilts, fantasies, hostilities, and conflicts which weigh them down and interfere with the assimilation of information and the decision-making process.

We propose that the genetic counseling situation resemble any other affectively significant (i.e., psychotherapeutic) counseling situation, and that the genetic counselor acquire the skills, knowledge, and attitudes needed by counselors in general. This kind of counseling is akin to that defined by two psychologists as a "way of life . . . in which facilitative interpersonal processes evaluating both counselor and client on the same core dimensions of interpersonal functioning occur." The core elements in their definition of counseling are: "empathetic understanding, respect or positive regard, facilitative genuineness, personally relevant concreteness or specificity, immediacy, or dealing with the here and now, and the ability of the client for self-exploration." [17] The value of these elements in facilitating counseling has been operationally defined and widely tested. Under this model a genetic counselor would be not only a trained geneticist but also "a person acutely aware of his own experiences" who conveys his respect for the client's strength to make his own decisions. The counselor's role is to elicit alternatives from the client, to help him distinguish his assets and liabilities, to enable the client sort out what he thinks he should or should not do, to support him in his decision or nondecision, and to make it clear that, as counselor, he is ready to give his support again. In this model, the counselor would be concrete, moving the counseling beyond empathy. He would be trained to understand and distinguish the fact, inference, and value aspects which are involved in any human transaction, and he would help the client to clear his path in order to reach the decision which is best for him.

We construe the role of the genetic counselor to be that of a facilitator to an important degree. Ideally, he should be sensitive to the hidden obstacles which make it difficult or impossible for the client to assimilate, to use, or even to hear (i.e., receive) the information that is being communicated. His training should provide him with sufficient awareness of his own biases and prejudices, so that he can recognize the points at which personal values may impair his capacity to be maximally helpful to the client. His own needs for power and dominance should be sufficiently worked through so that he will not, without wishing to do so, reinforce the client's tendency to be dependent upon him. He needs to have sufficient psychologic sophistication to recognize the transference process which makes the client accept his advice uncritically (or rebel against it), or to be aware of his own tendencies to identify with a client, which may interfere with the detachment necessary for a collaborative examination of the family's future.

To ensure maximum respect for individual autonomy, the genetic counselor should thus adopt a nonauthoritative, egalitarian stance, not so much giving "knowledge" as enabling counselees to make choices based on that knowledge.

In redefining the relationship of genetic counseling to more generic counseling, we are stripping it of many of the aspects which have formerly placed it within the medical model. By this, we do not mean that physicians should not counsel. Effective counseling need not be different from other communication. While certain external criteria do differentiate genetic counseling from generic counseling, these differences should not alter the general manner in which the counselor relates to the counselee. The generic approach is an eclectic one, fraught with as many or more of the difficulties of traditional genetic counseling. This view of genetic counseling still embodies interactions between those designated by society as "more knowing" and "less knowing." In presenting such a model, we are consciously advocating the primacy of the individual. The only way to make such a model work, we believe, is to base it on *what has greatest effectiveness for the client and his perceived needs*.

This orientation is rooted in our basic premise that at this stage of our knowledge of genetics, the rights of the individual transcend those of society, a position to which some (though by no means all) genetic counselors, officers of genetic societies, and international bodies adhere.[10,18,19] It is possible (but not necessarily true) that such a posture presents the least threat to individual freedom of choice, and hence the least possibility of restricting autonomy. (We must also point out that in advocating autonomy, we are also rejecting "coercion" as universally undesirable—itself a value judgment.)

Independent of the strength of the counselor's ideologic commitment, however, the potential for controlling behavior will continue to exist in any counseling relationship. The counselor's prestige, his conferred expertise, the conditions of great personal distress frequently brought to counseling, and the awesome medical milieu, coupled with the client's frequent lack of knowledge about his condition and the often enormous sociocultural gap between the professional and his client, combine to reinforce the client's vulnerability and dependency, and thus his capacity for being manipulated.

We believe the counselor's minimum obligation is to analyze the nature of his potential for exercising psychologic control over those individuals who come for counseling. This analysis may utilize a psychodynamic reference frame, with special attention to the concept of transference-countertransference; an operant frame, with attention to the counselor's unconscious reinforcement of particular client behaviors; or a transactional, interactional, existential, or other frame of reference.

We believe that the success of counseling can then be defined only in light of whether the process has facilitated or impeded the counselee's ability to make his *own* decision. The "rightness" of that decision can never truly be ascertained.

Such an analysis of a genetic counselor's approach to counseling presents difficulties, yet it seems essential if the counselor is to recognize his own value constructs as integral parts of the help he offers. It presents him with the opportunity to make *implicit* controlling factors *explicit,* thereby making the counseling process more conducive to the ideal of counselee autonomy.

In summary, because of the fundamental emotional conflicts (involving the individual, family, and society) inherent in carrying and transmitting genetic disease, genetic counseling can (and, we believe, should) be approached as an instance of all affectively significant psychotherapeutic counseling. Therefore, we believe the genetic counselor must have the same skills and attitudes that the generic counselor has.

Our incomplete knowledge of prevention and treatment of genetic disease (particularly its long-term impact on society, and on the gene pool itself), coupled with the potential threats to the client's psychologic freedom inherent in directive counseling, suggests that the practice of genetic counseling will usually require a nondirective and facilitative approach. When genetic counseling is not understood as an instance of counseling in general and is not done in a nondirective manner, we believe it has the potential of transgressing the bounds of professional service and may best be understood as "behavior control" in its pejorative sense. We believe that every counselor has an obligation to examine the morality of the use of psychologic power to control the behavior of any counselee in his charge, and to ascertain if that same power might better be used to help the person maintain his own autonomy.

REFERENCES

1. Gaylin, W.: On the borders of persuasion—a psychoanalytic look at coercion. *Psychiatry*. Feb. 1974, pp. 1–9.

2. Murray, R. Jr.: Problems behind the promise: ethical issues in mass genetic screening. *Hastings Center Report 2* (2):10, 1972.

3. Valentine, G. H.: *The Chromosome Disorders*. Philadelphia: J. P. Lippincott, 1969, pp. 153–155.

4. Veatch, R. M.: Value-freedom in science and technology. Ph.D. dissertation, Harvard University, 1969.

5. Durkheim, E. *Suicide*. Spaulding, J. A. and Simpson, G. (trans.). Glencoe, Ill.: Free Press, 1951.

6. Sorenson, J. R.: Factors shaping decision making in applied human genetics. Unpublished manuscript. Princeton University, 1972, pp. 19, 20, 7.

7. Solnit, A. J. and Stark, M. S.: Mourning and the birth of a defective child. In *The Psychoanalytic Study of the Child*. New York: International Universities Press, 1961, vol. 16, p. 523.

8. Lax, R. F.: Some aspects of the interaction between mother and impaired child: mother's narcissistic trauma. *Int. J. Psychoanal. 53:*339–341, 1972.

9. Langsley, D. G.: Psychology of a doomed family. *Am. J. Psychother. 15:*531, 1961.

10. Lappé, M.: Allegiances of human geneticists: a preliminary typology. *Hastings Center Studies 1* (2):63–78, 1973.

11. Reed, S.: *Counseling in Medical Genetics,* 2nd ed. Philadelphia: W. B. Saunders, 1963.

12. Fraser, F. C.: Genetic counseling and the physician. *Can. Med. Asso. J. 99:*19, 1968.

13. Carter, C.: Discussion. In Berg, J. M. (ed.): *Genetic Counselling in Relation to Mental Retardation*. Elmsford, N.Y.: Pergamon Press, 1971.

14. Veatch, R. M.: Ethical issues in genetics. In Steinberg, A. G. and Bearn, A. G. (eds.): *Progress in Medical Genetics,* New York: Grune & Stratton, 1974, vol. 10, pp. 223–264.

15. Fletcher, J.: The brink: the parent-child bond in the genetic revolution. *Theological Studies 33:*457, 1972.

16. Reiss, J. A. and Menashe, V.: Genetic counseling and congenital heart disease. *J. Pediatr. 80:*655, 1972.

17. Carkhuff, R. R. and Berenson, B. G.: *Beyond Counseling and Therapy*. New York: Holt, Rinehart and Winston, 1967.

18. Dobzhansky, T., Kirk, D., Duncan, O. D., and Bajema, C.: *The American Eugenics Society, Inc. Six Year Report*. New York: American Eugenics Society, 1970.

19. *Genetic Counseling. Third Report of the WHO Expert Committee on Human Genetics*. World Health Organization, Technical Report Series, no. 416. New York: American Public Health Association, 1969, vol. 13.

9

THE IMPLICATIONS OF SHARING GENETIC INFORMATION *

Stanley Walzer, M.D.,
Julius B. Richmond, M.D., and
Park S. Gerald, M.D.

The purpose of genetic counseling is to advise prospective parents about the risk of having a child who may be afflicted with a congenital or hereditary disorder, and to help them deal with the feelings that develop as a consequence of receiving this information. Often the parents have had a relative or another child in whom an abnormality has been clinically diagnosed, and they want to know what the chances are that their future children will be similarly affected.

More recently, as it has become possible to diagnose carrier states with greater accuracy, subtler problems have arisen; the prospective parents have had to arrive at judgments concerning risk even when a family member has not been demonstrably affected. With the introduction of biochemical and chromosomal screening of newborn infants, the problem has become even more complex. For example, genetic infor-

* Work for this chapter was completed while one of the authors (S.W.) was supported by Research Scientist Development Award MH–25070 and USPHS National Institute of Mental Health Research Grant MH–17960.

mation must be shared with parents of phenotypically normal infants (i.e., balanced rearrangement carriers) or with parents of infants whose chromosomal variations—although "silent" at present—may have implications for development in later life, i.e., XXY, XYY, etc. Questions are being raised about the advisability of sharing such information because of the impact it might have on child-rearing practices.

In this era of interest in basic developmental processes, the demand for genetic advice has continued to grow, as both the public and medical practitioners have become better informed and as the number of clinical geneticists has increased. Frequently, however, genetic counseling has involved a very short-term contact with a family, with the follow-up being done by the family practitioner or not at all. Thus, the genetic counselor rarely sees the long-term impact of such information on his clients.

We are currently involved in a long-term developmental study of infants with chromosomal variations. This report will not consider the primary research aim of our behavioral study—to investigate the nature of the relationship between sex-chromosome variations in male neonates and subsequent personality development. We will, however, discuss the effect such genetic information has had on the families concerned, since this issue has been a complex and critical one in the study.

Methods

Neonates with chromosomal variations have been ascertained through a chromosomal screening of newborns, utilizing heel-stick blood, a micromethod for peripheral blood culture, and microscopic karyotyping to establish the chromosomal complement.

During a four-and-a-half year period, 13,751 neonates were successfully karyotyped. Major chromosomal abnormalities were ascertained with a frequency of 6 per 1,000 neonates successfully cultured.

The infants with XXY and XYY karyotypes are being observed in a long-term developmental study in which systematic contacts are maintained with the family. A specific number of infants with familial balanced autosomal rearrangements are also being observed in the developmental study as one control population.

The families of all the infants with chromosomal variations have been informed about the abnormality and have had extensive genetic evaluations. Our sample may be divided into three groups:

(1) Phenotypically abnormal neonates ascertained at birth to have a chromosomal abnormality (Down syndrome, trisomy 18, etc).
(2) Phenotypically normal neonates ascertained at birth to have a chromosomal variation (i.e., balanced autosomal rearrangement or a metacentric Y chromosome). The families were informed, evaluated, and counseled *but not followed*.
(3) Phenotypically normal neonates ascertained at birth to have a chromosomal variation (i.e., XXY, XYY, and a specific number of balanced autosomal rearrangement carriers). The families were informed, evaluated, and counseled; and they *are being followed* in the long-term developmental evaluation.

For the purposes of this report, we will not present data on the phenotypically abnormal children with chromosomal aberrations (i.e., Down syndrome and D_1 trisomy). Rather, we will focus on the instances of phenotypically normal children whose chromosomal abnormalities were unexpectedly ascertained in the newborn survey.

Informing the Parents

Recently, both the public and professional communities have become increasingly concerned about what prospective subjects are told prior to any screening procedure. There seems to be consensus that families be sufficiently informed about its significance, implications, risks, and benefits to allow them to make an informed and independent decision about participation or nonparticipation in the screening procedure.

Prior to the chromosomal screening in our study, the prospective parents were informed about the screening procedure and were told that they would be notified if any significant chromosomal aberration was found in the infant. Therefore, both ethical and legal factors determined that we inform the parents about any significant variation ascertained and provide them with the established facts about its developmental relevance.

During the last two years, the informed consent procedure employed for this chromosomal screening was modified twice in order to ensure that more specific information about the chromosomal screening and developmental study was available to the prospective parents. More information was given about the frequency of occurrence and the nature of the major chromosomal abnormalities that could be ascertained. Prospective parents were also given information about the developmental study and about the risks and benefits associated with the screening and follow-up procedures.

If the neonate was phenotypically normal at birth, we usually waited until he was four months old before contacting the family. This gave the mother an anxiety-free period in which to get to know her infant; she was then better able to accept reassurances as to the developmental irrelevance or limited relevance of the variation.

Since ascertainment bias so colors what we know about a variety of chromosomal disorders, the information we shared had to be carefully considered. For example, it has been estimated in the past that from 20 to 40 percent of infants with the XXY karyotype have some degree of intellectual retardation. However, DQ (developmental quotient) scores of our XXY and our control children were similarly distributed (Table 1). No child with mental retardation (i.e., DQ below 70) has been ascertained.

It must be emphasized that the DQ test during infancy is a relatively poor predictor of the IQ in later life for normal infants. If the infant is retarded, however, his future IQ score can be more readily predicted. Thus, although the distribution of IQ scores among the four categories shown in Table 1 will change, it is evident that very few—if any—of these children will end up in the retarded range. Epidemiologic studies suggest that some risk for retardation might exist with the XXY karyotype, but the degree of risk quoted by some textbooks is surely too high.

We informed the parents of the infants with an XXY karyotype that testicular changes and delayed pubescence were possible phenotypic

TABLE 1
Mental Developmental Indices
(Bayley Test)

XXY			
Dull Normal (80–89)	Average (90–109)	Bright Normal (110–119)	Superior (120+)
85	93	110	122
	100	110	127
	105	112	132
	107	112	
		115	

Controls			
Dull Normal (80–89)	Average (90–109)	Bright Normal (110–119)	Superior (120+)
	103	110	128
	107	110	
		117	
		118	

variations that could appear in individuals with this chromosomal complement. With the XYY karyotype, tallness was mentioned as a possible phenotypic attribute. We explained—in terms that parents could understand—the limitations that existed in our knowledge about the behavioral implications of these karyotypes as a result of the ascertainment bias which so colors this information. We further explained that a more precise assessment of any special hazards that might be associated with the presence of these extra chromosomes would result from prospective studies of children identified at birth as having these variations.

When an infant is ascertained to have a structural rearrangement, one parent almost always carries the same rearrangement; therefore, this family might be at risk with respect to future pregnancies. The appearance of the rearrangement in its imbalanced state could result in progeny with congenital abnormalities or retardation. Locating chromosomally high-risk parents permits us to perform early amniocentesis (in certain cases) in subsequent pregnancies to establish the karyotype of the fetus. Should a serious chromosomal imbalance be found—one whose presence is known to cause difficulty—the possibility of surgical intervention to terminate the pregnancy could be considered. It has been our firm policy, therefore, to inform these parents fully about the chromosomal abnormality ascertained in their child, to evaluate the family genetically, and to explain the risk of having abnormal children in the future, should such risk exist.

Initial Contact and Parental Response

When the infant was four months of age, the parents were informed about the variation uncovered. In the earlier phases of the study this was done by telephone; however, parental concern generated a variety of questions that could not be answered effectively on the phone. Throughout most of the study the initial contact was made in a letter reporting the presence of a chromosomal variation and inviting the parents to meet with the genetic counselor. The immediate response of most parents was a concerned phone call to their pediatrician, in which they asked about the variation and—at times—expressed annoyance at not having been told earlier. We have therefore informed the pediatrician about the chromosomal variation (in general terms) prior to sending the letter. We have encouraged him to reassure the family about the limited developmental relevance of the finding, leaving the specific explanations to the genetic counselor at the time of the evaluation. Most of the families accepted the reassurance offered by their pediatricians but did request immediate information about the aberration. Therefore, they were seen within 24 hours after receiving the letter.

The vast majority of the families moved with relative comfort into the personal meetings with the genetic counselor. For a few families, however, the short period between the initial contact by letter and the personal meeting was a difficult one. In one family, the mother was physically ill and—unknown to us—was hospitalized at the time of our initial contact. The father insisted that no one inform her until she recovered. However, within an hour we received a concerned phone call from the mother in the hospital. The family had felt the need to share with her immediately their concern about the variation.

In another family the father requested a complete, formal, and written report before agreeing to see us.

In a third family, in which the infant had an autosomal rearrangement, the father politely but firmly informed us that the infant was well and that nothing was wrong; he preferred that no further consideration be given to our finding.

The meetings with the parents were used not only to impart knowledge and answer questions about the variation but also to deal with the feelings associated with receiving such information. The counselor was particularly careful to identify the psychologic processes employed by the parents to cope with the information, appreciating that these processes, in particular, would significantly color what the parents "heard" and retained.

We were impressed by how frequently the psychologic mechanism of denial was employed initially to cope with the information imparted. This mechanism was difficult to identify, since the give-and-take transaction between the parents and the counselor left each participant feeling satisfied with the contacts. The parents not only seemed to gain an understanding of the chromosomal variation but also appeared relatively comfortable with the information shared with them. The counselor, in turn, felt satisfied that he had not simply "dumped" genetic information onto individuals ill-equipped to receive it.

This is the stage at which many clinical geneticists would lose contact with a family, feeling confident that all was well understood and assimilated. However, long-term follow-up of some of these families has permitted us to assess the full impact of the information imparted to them.

Long-Term Parental Responses

We have continued our direct contact with those families included in the long-term developmental study. The patterns of parental responses that have been seen must be attributed both to the continued reaction to the information imparted and to feelings about being included in a long-term behavioral study.

The parental response to being informed about the existence of a chromosomal aberration in their child did not appear to be related to the amount of information given prior to screening. We noted the same parental concerns and coping strategies before and after the modifications in the informed consent procedures employed prior to screening.

After the initial meetings with the counselor, many of the families appeared at ease with the information they had received and settled into a productive relationship with the developmental study. However, a few other parents gradually developed second thoughts about the chromosomal aberration, again feeling concern about its origin and the developmental implications it had for their infant. The "silent" nature of the chromosomal aberrations (i.e., XXY, XYY)—with the phenotypically normal infant carrying a variation that could not be seen but conceivably might have some developmental significance in the future—gave these parents a vehicle through which the realistic information previously shared with them could become colored by their own personal concerns. The parents felt the need to "test" the information given them by sharing it with their own pediatrician; they would report that they did not fully understand all that was explained and would like to hear it again from their own physician.

During the first visit for the developmental study, these parents asked questions about information previously "understood" by them. These questions were colored by fantasized developmental implications that were in no way related to the aberration. One mother mentioned that a neighbor had a child born with normal intelligence who "developed retardation" during his childhood years. Since the neighbor had emphasized that this was "inherited," the mother wondered whether this could happen to her child. During the next visit, the mother informed the investigator that no such neighbor existed but she simply "wanted to see if such a thing could happen."

It is of interest that the parents' education or socioeconomic level did not seem to influence their responses to receiving genetic information and being included in the study. Some of the more complex situations we had to deal with involved the better educated, more affluent families.

It appeared to us that the intellectual issues became clouded by the parents' own concerns, and it was this admixture that had to be coped with over time. Furthermore, it was evident that some of the "intellectual understanding" that the parents demonstrated in their meetings with the genetic counselor resulted—in part—from their use of the psychologic mechanism of denial. Since this mechanism interfered with the process of fully assimilating the information given them, the gradual realization that their infant did indeed carry a "silent" variation left them with concern that had to be coped with.

Genetic issues periodically returned to the foreground. The parents often came into contact with articles, as well as with TV and radio programs, on a variety of genetic subjects. For a short period after they received the genetic information on their own infant, some parents would tend to apply what they read or heard to themselves, even though this information usually was in no way related to their child's variation. Thus, during the early visits for the follow-up study, the investigator had to reassure parents that their child's aberration was not related to the genetic diseases currently being publicized through the mass media.

With continued collaborative discussion, explanation, and support, almost all of the families have settled into a long-term relationship with the study that might be described as one of "constructive involvement." Many of the family members have sought the advice of the investigator (S.W.) on a wide variety of family concerns, including marital stresses, illnesses, economic problems, educational questions, and developmental issues relative to the affected child or others. The parents often prepare for the periodic home visits, presenting excellent and carefully thought out developmental reports about their child. The families have kept in close touch with the study and have frequently verbalized the wish to remain associated with it. If they moved from the city, they arranged to have their child continue in the study.

Most of the families have developed a high level of comfort with the information about their child's chromosomal aberration. Any questions or concerns were verbalized directly and openly, and the information supplied was integrated with ease. The questions became fewer and more realistic with time. In many instances, the parents would ask the question and then answer it themselves, having available the basic information and the capacity to sift out the fantasized concerns from the realistic knowledge.

Many of the parents became able to analyze genetic articles they encountered in the press with a higher level of critical competence than was demonstrated by some members of the professional community. They could correctly interpret statements and assumptions based on information colored by ascertainment bias. For example, one parent who had read a short article on the XXY karyotype and its alleged relationship with mental retardation commented: "This guy [the author] is still stuck back at the time they got their knowledge by screening in hospitals for the mentally retarded, where everybody tested was retarded. What about the knowledge gotten from people like us?"—meaning the results obtained from screening populations not selected because of retardation.

Time Takes Its Toll

The outcome described above was the result of the investigator's being available to the families long after the information about the abnormality was imparted to them. What happened in instances in which genetic information was imparted but the families were not followed?

Fortunately, we had an opportunity to contact several of the families in which a chromosomal lesion had been ascertained (i.e., the infant and usually one parent had a balanced autosomal rearrangement); they had met personally with the counselor but had not been followed in the developmental study.

During the last four years, when banding techniques for individual chromosome identification became available, we again contacted 12 of the families that had not been seen since the early life of the propositus, in order to ask consent for and obtain new blood samples for fluorescence studies. The interval between the original contact and the second visit ranged from 1½ to 4 years.

We were surprised to find how much of the information we previously presented to the families was forgotten; this caused us to wonder about the advisability of informing about chromosomal variations without follow-up contacts. Although many of the families remembered that "heredity" was the issue, 5 of the 12 could not remember that chromosomes were involved; 2 others remembered this fact but could not identify the nature of the variation. All 5 who had retained no information about the chromosomal nature of the abnormality did remember that it had something to do with an aberration diagnosed through an earlier blood test. One parent asked: "Have you found other children who had this and were slow?" Another made the statement: "We never understood it anyway so we just forgot about it."

If other investigators had studied these families at this point, they would have been impressed by the families' lack of knowledge and would have raised questions about the information given them at the time of the evaluation. It was evident that these families had not retained the facts during the intervening period.

One family contacted us years later, wondering whether we wanted to see their child. They wanted to show us how well he had developed, emphasizing that we would be "surprised to know about this," as though we had predicted otherwise. In fact, we had given them much support and had emphasized the developmental irrelevance of this variation. In our visits with this family, the child appeared to be happy and healthy, developing well in all significant respects.

Discussion

The wide exposure given to genetic issues in the mass media, along with the variety of speculations about the potentialities for genetic manipulation, has endowed genetics with a mystical and somewhat fearful "power" to shape our lives. It is clear that such accounts have contributed to the confusion and concerns manifested by several of our families.

Genetic counseling introduces difficult concepts that are not easily assimilated by the families being counseled. They seem to understand at the time of the evaluation, and perhaps have no further questions, doubts, or concerns. Doubts continue to be raised, however, as the original denial wears off and they learn of genetic issues from the media, their friends, or their physician. The facts that originally are presented to them are colored by their own concerns and misinterpretations. During this period they turn to their local practitioner, who may be only vaguely informed about the implications of chromosomal aberrations. It is obvious that if genetic information is to be shared, the relationship with the genetic counselor must be maintained so that questions and concerns arising later can be answered.

Other studies have emphasized the difficulty that families have in understanding and retaining information about congenital abnormalities in their children. For example, Reiss and Menashe [1] assessed the ability of a group of parents to recall the recurrence risk of their child's heart defect. Of the 39 families interviewed, none were aware of the recurrence risk; only 9 of the 39 knew the potential mathematical risk when questioned 1–4 months after being told it again.

Our study has emphasized that the most relevant issue is not the sharing of genetic information. If the counselor is sensitive to the psychologic state and needs of his clients, and if he can be available from time to time to answer questions and allay doubts as they arise, the parents can cope with the information in a way that does not disrupt their child's upbringing. The counselor should not give the family genetic information with the assumption that it will be fully understood and retained unchanged over time.

Investigators concerned primarily with the ethical aspects of chromosomal surveys emphasize the dilemmas arising from the need to meet the demands for informed consent. Parents have the legal and moral right to know the facts about their child's abnormality. On the other hand, conveying such information could create concern in the parents about their child's future development. The question is raised: Would it not be better if the chromosome test were not done at all, so that no one would be affected by the disturbing knowledge?

The answer to this question lies in the answer to another question: What is the risk to the child when no knowledge of the variation exists? If the sex-chromosome aneuploidy states in males are related to behavioral and/or physical abnormalities that might be amenable to early therapeutic intervention, then it is important to recognize these conditions early in the child's life. Preliminary data on the XXY (and perhaps on the XYY karyotype) suggest that the affected individuals may have some potential to exhibit subsequent developmental handicaps or psychopathologic problems. However, definitive knowledge about the actual degree or nature of the risks and the success of early intervention strategies for modifying these risks depend on long-term prospective studies of affected individuals.

The negative aspects of genetics color even the attitudes of clinicians and investigators close to the fields of clinical genetics and psychiatry. We are frequently asked this question: Why bother doing chromosomal screenings on newborns and imparting potentially traumatic information to families when you cannot do anything to help the affected individuals anyway? Such pessimism must be examined with respect to behavioral development. To predict with certainty the exact behavioral outcome of a particular genotype is to deny the most basic concept of development; the genotype expresses itself only in interaction with the environment at all levels of development—cellular, fetal, neonatal, and postnatal. This is as true for behavioral development as it is for physical development.

Certain genotypic variations might indeed narrow somewhat the variability possible in certain areas of behavioral development. The more serious the genotypic abnormality, the more limited is the variability possible in specific developmental areas; if a male carries two or three extra X chromosomes rather than one extra, he is much more likely to manifest significant mental retardation. The more severe the environmental stress, the more limited are the developmental options for the unfolding genotype; thus, complete starvation to the point of death allows no options for development, since the genetic material ceases to exist. However, within the limits imposed by these extremes, there remains variability in the phenotypic expression of the genotype. With respect to behavioral development, without the concept of variability one is left with a simplistic, mechanistic, reductionistic biologic psychology—a "homunculus"-oriented biopsychology.

If variability is emphasized, why are genetic studies of human behavior performed in the first place? Because we hope to be able to demonstrate whether or not the range of variability in particular areas of behavioral development is narrowed by specific genotypic abnormalities. This indeed can be recognized by careful investigation without denying the existence of variability.

Since the sociocultural environment is such a profound force in the genotype-environment interaction, modification of the environmental factors can have a significant effect on the unfolding genotype. Therefore, there is no reason to assume that therapeutic intervention strategies would not help to prevent some of the developmental handicaps that might be associated with the sex-chromosome aneuploidy states.

This can be demonstrated in our own study, where five of eight children with the XXY abnormality (over five years of age) have demonstrated speech and language defects. Early speech therapy with two of them resulted in improvement. The other three did not require speech therapy for their problem. Three of the XXY boys with language disorders and one without language problems went on to demonstrate early reading difficulties. The relationship between language disorders and learning difficulties in normal children has been well described in the psychologic literature. Early therapeutic tutoring in reading—before the problem became chronic—resulted in improvement in three of these boys. Perhaps the speech and language dysfunction and/or subsequent reading difficulty might contribute to the lack of academic success seen so frequently in young men with Klinefelter syndrome, even though they are of normal intelligence.

There are many such instances which emphasize the benefits of early knowledge. Whether it is or is not known does not change the fact that the child does have a chromosomal abnormality. If the abnormality presents a developmental risk that can possibly be modified, it is important to recognize the risk and treat the child. It is probable that appropriate psychologic, educational, or therapeutic intervention can influence the pathologic outcome of these defects. It is through long-term developmental studies combining systematic behavioral observations with therapeutic intervention that we will obtain the necessary information about the actual "behavioral risk" of these karyotypes.

Some critics of the policy of informing parents about the significant chromosome aberrations ascertained in their children employ the doctrine of the self-fulfilling prophecy to support their argument. According to this doctrine, the parents' concerns about or expectations of the possible development of a particular behavior or characteristic will result in child-rearing practices that actually foster the development of the feared behavior or characteristic. Thus, proponents of this doctrine feel that parental concerns about or expectations of mental retardation or antisocial behavior will contribute to the development of mental retardation or antisocial behavior in the child.

Bypassing entirely the issue of the scientific validity of the concept, if such a phenomenon did exist and were relevant for longitudinal studies such as ours, we would have expected a significant increase in mental retardation in the children with chromosomal aneuploidies. The possibility of mental retardation was the most commonly expressed concern among the parents of the aneuploid infants. Yet the DQ scores of our XXY and control children were similarly distributed (Table 1). There were no mentally retarded children in our study.

Valentine [2] believes that parents of XYY infants should not be informed about the presence of this aberration. In spite of his feelings, he had to inform three of the four families with XYY children about the variation, either because they had figured it out themselves or the child was to be adopted. The one family which had figured out the problem later expressed the wish that they had not known in the first place. However, the parents' knowledge about this variation apparently had little effect on their child's development. Valentine reported on the child as follows:

> At just two years of age he was seen again. The parents reported entirely normal development and behavior, stating that they could not see any difference in temperament from his brother. The mother admitted to anxieties about disobediences and rebellious behavior that, had she not known of his abnormality and its possible implications, would otherwise have caused her no concern. Nevertheless, both parents appeared to have adjusted well to the knowledge of their son's XYY constitution. His physical examination showed a handsome, happy, cooperative, and normal child of 13.61 kgs. and 84.5 cms.

In another case, Valentine again informed the parents about the XYY complement, without any adverse developmental effects being noted. This young infant was adopted by his uncle since his natural mother did not wish to keep him. Valentine reported: "These adopting parents were told in detail about the abnormal chromosomal complement and the state of our knowledge, such as it was, of its implication. The pediatrician who would now be looking after the baby was similarly informed." At 13 months of age the child's development was reported by his pediatrician as normal. Valentine reported a dictated personal letter from the adoptive mother which described the child as a "very happy, contented, easy-going and lovable baby who was no different in personality from her own three children at the same age." At 16 months of age, the baby was again described as being an entirely normal infant. Again, parental knowledge about the XYY karyotype apparently did little to affect the child's development adversely.

We have routinely asked each parent whether he or she would rather not have known about the chromosomal variation. Although several parents did verbalize their wish that the chromosomal variation had not been there in the first place, no parent has ever expressed the desire not to know about something that might have significance for the child's development.

The importance of obtaining data less colored by ascertainment bias is further emphasized by the demands now being put on clinical geneticists by the increasingly common procedure of amniocentesis. Genetic counselors and their patients must now make judgments about the advisability of interrupting a 20-week gestation if an XXY or an XYY karyotype is detected in utero. This decision is very difficult to make on the basis of currently available data.

As more genetic knowledge becomes available, it is likely that many more parents will wish to have access to it. Certainly it is the obligation of our society to expand our knowledge so that we can provide increasingly accurate information to those who seek it. Such information can be obtained through further studies which respect the rights and dignity of families while fulfilling our obligations to obtain and provide more knowledge. These studies should lead us further along the road to prevention, which is our ultimate goal.

REFERENCES

1. Reiss, J. A. and Menashe, V.: Genetic counseling and congenital heart disease. *J. Pediatr. 80:*655, 1972.

2. Valentine, M. B., McClelland, M. A., and Sergovich, F. R.: The growth and development of four XYY infants. *Pediatrics 48:*583, 1971.

10

ETHICAL ASPECTS OF GENETIC COUNSELING

Albert S. Moraczewski, O.P., Ph.D.

Introduction

The triad of psychiatry-genetics-ethics represents a formidable combination. The intersection of these three different lines of inquiry, each with its own methodology, presents a situation of bewildering complexity. It is only in recent times that genetics and psychiatry have looked at each other with any degree of comprehension. To introduce into that relationship a third element, ethics, may make the psychiatrist look at the whole enterprise with some apprehension. He may feel that psychiatry is sufficiently difficult without complicating it with both genetic and ethical considerations.

Following hard upon the rapid development of medical genetics, including such techniques as amniocentesis, there has been an accelerating expansion of genetic knowledge into medical practice. With increasing frequency, physicians are asked by patients to give advice concerning problems associated with genetic defects. Whether or not the psychiatrist is the best one to give such advice (this point has been raised by Hecht and Holmes), experience reveals that increasingly he will be sought to help the patient cope with his problems.[1] In this area, psychiatrists will be hard put to ignore ethical questions, since genetic problems carry along with them special ethical and moral issues. Some are more obvious, such as those related to the aborting of a fetus with a genetic defect; and some are more subtle, such as those associated with the effect of genetic information on the individual's self-image and on the husband-wife relationship. Unfortunately, there is at present no clear ethical consensus on these and related issues. One need only read the literature on ethical aspects of genetic counseling and genetic engineering to recognize the diversity of views on these topics. An example with significant implications for social policy is that of mass screening for genetic disease, in which we find at least two positions being argued. A research group from the Institute of Society, Ethics and the Life Sciences proposes a set of principles for guiding the operation of genetic screening programs which emphasize that the programs be voluntary.[2] Contrariwise, John A. Osmundsen, emphasizing the many benefits of genetic screening, recommends compulsory screening under certain conditions.[3] Notwithstanding these difficulties, we must admit along with Robert S. Morison that modern genetics has raised profound questions of value that must be faced.[4]

A "gut-level ethics" is not sufficient; merely to consult how one "feels" about a problem does not ensure that one's ethical decision will be sound. Such a decision should result from a rational analysis, which might well be considered an integral part of genetic counseling. Consequently, our overall objective is to indicate how to make ethically sound decisions in situations involving a problem related to genetics. To this end, it will be necessary to give some consideration to the ethical decision-making process itself. This chapter will be divided into two main sections. The first will consider very briefly the relevant aspects of ethics. The second will be concerned with the ethical dimensions of genetics and genetic counseling as they relate to psychiatric practice.

General Ethical Considerations

Before embarking on a specific discussion of ethics, I would like to make a few preliminary remarks indicating my ethical position. These days the ethicist, particularly the religious ethicist, is perhaps overly sensitive about placing restrictions on scientific endeavors in light of Galileo and Darwin, to mention but two notorious examples. He certainly does not wish to impede the legitimate progress of science, but he is also concerned about the rights and dignity of man in the face of an overwhelming technology. Some, I fear, would take the attitude that what technology *can* do, it *may* do. An ethicist of this persuasion would tend to bless all the efforts of science, seeing them as contributing unqualifiedly to the development of the human person and his society.[5] Other ethicists, like myself, feel that technology does not have carte blanche to do whatever it may wish to do. As Oppenheimer has observed, we shall have to learn that not everything which is technically sweet is morally good. A similar sentiment is offered by A. W. Diddle: "We undoubtedly have the potential to manufacture a human person. But should we?"[6]

The contrary position is that technologic might is moral right. An ethical pessimist would say that the "progress" of science and technology is inevitable; a man could sooner stop an avalanche with his bare hands than he could prevent indefinitely the technologic application of knowledge. This paper is based on the presumption that man can make effective ethical decisions. The atomic bomb need not have been dropped—or even constructed.

It is a commonplace but nonetheless agonizing observation that persons of comparable intelligence, knowledge, and good will are sharply divided ethically on vital issues such as abortion or war. But why? Tracy M. Sonneborn raises the question of who decides what is right and what is wrong.[7] In one sense, it is the individual who decides what is right and what is wrong in his own case. This decision process involves the application of ethical principles or guidelines to a particular moral problem.

It is this process which admits the introduction of differences among persons. Different sets of ethical principles can be employed, or individuals using the same set of principles may assess the particular problem differently. In addition, many factors, emotional, cultural, and social, influence the decision-making process. If the complexity of the situation were more widely appreciated, fewer accusations of callousness or ethical insensitivity would be hurled about.

Nevertheless, another aspect of ethical questions must be considered.

While we admit the powerful influence of many individual factors, we must recognize that some consensus on these issues is possible if open discourse is fostered. A difficulty in ethical discussions is what is left unsaid, what is assumed. But in a society like ours, with pluralistic values (for better or worse), it is of the utmost importance that nothing be presumed in such discussions. For example, "I am sure no one would wish to deny individuals the *right* [emphasis added] to decide not to have children who are genetically abnormal." [8] For a fundamentalist Christian who holds that all children come from God and that such natural misfortunes as genetic defects are to be patiently borne, that statement would not be acceptable. Furthermore, whence comes the alleged right? Is it rooted in civil law? Or does it have a more fundamental origin? Perhaps there is a societal consensus, albeit an unspoken one. I raise these issues here to emphasize the labyrinthine complexity of these questions. What do we mean when we say "I have a *right*"? What do we mean when we say that an action is right or wrong, good or evil? What is understood by the phrase "good is to be done and evil to be avoided"? There are some things that men do which we say are wrong, e.g., murder, stealing, lying. But why are these actions considered morally bad? It is the role of ethics to reflect on what makes those actions for which a man is personally responsible good or bad. The broad spectrum of meanings attached to the words "good" and "bad" accounts for the development of different ethical systems throughout the ages.

Rather than enter into a prolonged historic and philosophic analysis of these terms, for the sake of brevity and clarity I shall define them as they are used in this article.[9]

Right: a moral power vested in a person, owing to which the holder of the power may claim something as due him or belonging to him, or demand of others that they perform certain acts or abstain from them. By saying that *right* is a moral power, I mean to emphasize the fact that it does not depend on any kind of might or physical force. By the term *moral power* I also mean that to deprive one of a right would be an offense against justice. Traditionally, it has been customary to distinguish between positive or acquired rights, which have as their immediate source the state, and natural rights, also known as the rights of man, which are inherent in the human person and hence are ordinarily inalienable. This distinction is recognized in the Declaration of Independence, which specifies that there are certain rights which by their nature are inalienable: life, liberty, and the pursuit of happiness.

Good: that which is desirable; that which is capable of perfecting or completing something. A morally *good act* is one which brings one closer to one's transcendental goal, with due respect for the rights of others, especially one's ultimate goal; that which functions as or is according to what it is supposed to be. This latter judgment will obviously depend on the society in which one lives, should there be a consensus regarding what a good man should be.

Evil: the negation of good; the privation or absence of something that should be present. Hence a morally *bad act* is one which impedes a person from his goal, from attaining his life's destiny.

Our ethical analysis of genetic counseling should be examined in the context of the leading ethical systems currently prevalent in the United States. The systems can be categorized into three principal groups. The first is predominant, whereas the other two appear to be critical responses to it. The following material borrows heavily from a study made by Benedict Ashley, O.P. (unpublished notes).

I. The American success ethic
  (a) *Origin:* The so-called Protestant (Calvinist) work ethic modified, especially after the Civil War, by the development of our industrial society, and subsequently secularized.
  (b) *Scale of values:* Highest value is "success," understood as individual, competitive achievement measurable in largely quantitative, economic terms and evidenced by a good public image, for which humanistic values are seen as secondary adornments. This achievement is also projected as an identification with a form of nationalism which sees America as the best, most progressive country and the model for the rest of the world.
  (c) *Mode of thinking:* Pragmatic. The end justifies the means; no emphasis on harmony or hierarchy of values.

II. Ethics of compassion
  (a) *Origin:* Rousseauistic primitivism, supported by certain Christian ideas; nostalgia for the "American dream"; democracy and freedom seen as undermined by American success ethic.
  (b) *Scale of values:* Primary values are freedom, sincerity, "natural human feelings," democracy, human equality.
  (c) *Mode of thinking:* Emphasis not on reason system but on sensitivity, compassion, frankness of expression. (The so-called counterculture is merely a new form of this mode of thinking.)

III. Ethics of science and technology
  (a) *Origin:* "Yankee know-how" and industrial experience, combined with the scientific ideal of the German research universities; became prominent in the United States only after 1880.

(b) *Scale of values:* Primary value is "truth" in the sense of progressive scientific research (often confused with naked fact), but with a strong emphasis on technologic application; conviction that human problems can be solved by social engineering.
(c) *Mode of thinking:* Denial of relevance of values; once value problems are reduced to technologic problems, then the "scientific method" can be applied.

The ethical system of a reflective person ultimately flows from his world view, and more specifically depends upon what he views man to be. Charles Reich's *Greening of America,* for example, describes three ways of looking at man as Consciousnesses I, II, and III—namely, goal-oriented individualistic man, organization man, and liberated man.[10] These three correspond roughly to the systems described above. Thus, the goal-oriented individualistic man is related to the American success ethic; organization man to the ethics of science and technology; and liberated man to the ethics of compassion. Generally, then, when reference is made to an ethical issue in the literature, it will be in terms of one of these three ethical systems. Which one is used in a particular case can make a significant difference in the outcome.

Before entering into the ethical aspects of genetic counseling, one more preliminary consideration should be mentioned, i.e., the ethical decision *process*. A great deal can be written on this topic, but again, for the sake of brevity, it will be sufficient to outline three modes of ethical decision making. Just as the value system can make a difference, so can the decision process markedly influence the ethical analysis in determining whether an action is good or evil.

I. Mode of Legality
   (a) A *code* of law is assumed as given (by God, by society, or even by oneself). This is accepted not because it is reasonable but because it is "authoritative" (voluntarism).
   (b) This code is then applied to particular situations by legal reasoning or "casuistry."
II. Mode of Reason
   (a) Principles of conduct are discovered (personally or by experts, perhaps with confirmation by authorities) by induction from experience, based on the consequences of alternative modes of action.
   (b) These principles are applied to concrete cases by a reasoning pro-

cess that seeks to discover the optimum advantage in a given situation.

III. Mode of Authenticity
  (a) Motives of action are uncovered, so that a person has insight into his deepest needs and feelings.
  (b) In particular cases, the person seeks a way of acting which is harmonious with these deepest needs.

An example of the first type, the mode of legality, is an ethical problem solved in terms of the Ten Commandments (e.g., "Thou shalt not kill"). This law, part of the Judeo-Christian tradition, has been interpreted to mean that the killing of an individual person innocent of any crime is wrong, that it is a moral evil. Are there any exceptions? This is one point which divides many who otherwise accept the Ten Commandments as a source of moral guidance. Those who opt for an absolutist position see that once an exception is made, another is found, and another, and so on until the moral guidance of such laws is left entirely to human caprice. On the other hand, those who wish to make exceptions to all moral principles see the absolutist's position as destructive of human freedom, as eventually dehumanizing and destroying what is best in man. Therein lies a dilemma for our age.

The second mode of ethical decision making, the mode of reason, proceeds from principles usually arising out of experience. This mode can be illustrated by the matter of privileged communication, such as that between doctor and patient, attorney and client, priest and penitent. Experience has shown that if the professional person involved may not retain such confidential information, most persons would be reluctant to seek the help they need and society would be the worse for it. Consequently, it would be considered unethical for a doctor, attorney, or minister to convey such information to others without, at least, a proportionately serious reason and explicit consent.

The mode of authenticity as a type of ethical decision making is exemplified by the notion of honesty: "Above all to thine own self be true." Such a person acts so that what he does reflects how he feels at the time. There is no room for mental reservations, for equivocation. The authentic person is rigorous about being open, even at the expense of insulting others. Compromise is not part of his vocabulary. Should he conclude that as a lover of peace and people he should not continue to support a war through taxation, he would decide not to pay his income tax.

It will have been noted that these three modes are not mutually exclusive. In real life, an individual is rarely fully consistent in the mode of decision making he employs to solve the daily ethical problems he faces. Yet, for the most part, he will tend to favor one over the others. All the above having been stated, we can now turn our attention to the specific ethical issues that the psychiatrist may meet in the area of genetic counseling.

Genetic Counseling: Ethical Dimensions

Since in the present context the genetic counselor is a physician, there is a basic question of the precise relationship between the physician and the patient. In medicine we often speak about the physician-patient relationship but the nature of that relationship is, in some respects, unspecified and obscure. For example, is the relationship one of father to child? Is it one of superior to inferior, or is it one of brother to brother? How one answers that question will determine to a great extent one's response to some of the subsequent issues. While it is true that the physician in his area of knowledge has more information and relevant experience than the patient, does that create a teacher-student relationship? Is that the implicit contract? Another possibility is that the relationship is one in which the patient hires the physician as his agent to execute certain treatments, much as he would hire an attorney to be his agent with regard to certain legal matters. If, then, the physician is the patient's agent, he cannot do any other than what the patient desires him to do. That may seem fairly straightforward. However, the question then arises, can the physician do something which he knows would be inimical to the patient? Or can he do something beneficial for the patient without the latter's specific request? Clearly, the safe presumption is that the patient has hired the physician for his own continued well-being or its restoration. If we accept then, at least provisionally, the model of the physician as the patient's personal agent, this relationship has important implications for discussion of ethical issues. It implies that the physician will always respect the ethical stance of his patient: obvious, but no less essential.

ABORTION

Of the numerous ethical problems associated with genetic counseling, undoubtedly the most difficult and persistent is that of abortion. At the outset, therefore, I wish to tackle this question directly. I do not expect that this discussion will change anyone's conviction or position with regard to this controversial topic, but I hope that it will at least focus the problem precisely.

Stating the issue in its simplest terms, some people consider abortion murder, and others do not. The former see the fetus as a human person. To the latter the fetus is merely an organism that is on its way to becoming a person but has not yet actually attained that state. Consequently, the fetus, being less than a person, does not have the same rights as its mother or other human beings have. Hence, in a conflict of interests, the existing person wins over the purely potential person.

The abortion controversy appears, then, to pivot around the critical question of what constitutes a person. Indeed, many ethical problems depend for their solution upon the concept of personhood. For this reason I will discuss the matter at greater length than is customary in the current genetic literature. If we turn to the dictionary for a definition, we find that the word "person" means a human being, a human as distinguished from an animal or a thing. These definitions of person are not adequate, for they seem to presume that the terms "person" and "human being" are fully congruent. Yet that is the very issue at stake. Do these terms refer to one and the same reality? A symposium on "The Beginnings of Personhood" was recently sponsored by the Texas Medical Center's Institute of Religion and Human Development. While no consensus was reached by members of this symposium, certain clarifications were made. First, the notion of person was recognized as something associated exclusively with the human race; that is, for a being to be considered a person, he must belong to the species *Homo sapiens*. Second, it was recognized that it is the person who is the subject of human rights. In this connection, it might be helpful to consider a definition of person proposed by the fifth-century philosopher Boethius and interpreted 800 years later by Thomas Aquinas. They defined a person as "an individual substance of a rational nature." [11] By the term "individual substance" is meant a complete, existing entity which is essentially separate from all other things and subsisting of itself; for example, it does not share another's color, shape, organ, or limb. The term "rational nature" means that this individual substance has the power to reason, to engage in abstract thought. Consequently, that which we call person cannot be shared by another being. While it is true that two persons who love one another do share many things of common interest, they do not share—much as they might wish to—their personhood, properly speaking. It cannot be handed over as a package, for it is not a thing apart from the total human being.

It is also necessary to distinguish carefully between the terms "person" and "personality." The word "person" as used in this discussion and as defined above is a metaphysic or ontologic term, while "personality" is more in the psychologic domain. There is, therefore, a difference between person, or ontologic person, and psychologic personality. "Person" refers to the entire human organism seen in its unity and as the basis of all its action, the source of its functioning integrity. Psychologic "personality" refers primarily to the manifestation or to the measurable aspects of personhood. It is that which can be observed of one human being by another. The psychiatrist and psychologist as such are primarily concerned with the notion of personality rather than personhood as it is used here. Part of the confusion associated with such discussions about personhood is probably due to the failure to define clearly the various meanings of person and personality.

Bearing the above points in mind, then, is the fetus a person? Does it have the attribute of personhood described above? We know what a thing is by the way it behaves. If we want to know what this unknown seed is, we bury it in some soil, water it, and observe what it becomes: an oak tree, a corn plant, a peony, or whatever. The human fetus is already an individual substance, but does it have a rational nature? Barring an accident, the human fetus will develop into a being which will be capable of engaging in abstract reasoning. This capability is a reality, albeit largely potential, in the fetus. It is the potentiality which makes a difference. Infrahuman fetuses do not have that potentiality. No matter what happens subsequently, they do not manifest the distinctively human quality of abstract reasoning, even granting that some chimpanzees have been trained by humans in the use of sign language in a manner which asymptotically approximates human language.

The human fetus, then, is seen as a person because (1) personhood is identified with that organism, initially a zygote with 23 pairs of chromosomes, which results from human copulation; and (2) it develops gradually without any discontinuity into the mature adult human. If personhood were not an attribute from the very beginning, then its appearance at a later stage of development would require that it somehow be inserted by an extrinsic factor. There is no evidence of such a factor. It is true, of course, that environmental and social factors are of fundamental importance for the development of personality as previously defined. It would be erroneous to consider a person as if it were a thing, like an organ or a limb. Rather, it is the whole organism viewed as a functioning unity; not just any organism, but one belonging to the species *Homo sapiens*.

Most of us, I dare say, have only a rather static picture of the developing fetus. A. W. Liley, in a recent article, has described fetal activity very graphically. Because it effectively conveys the dynamic individuality of the fetus, I shall quote at some length from this article.

> Far from being an inert passenger in a pregnant mother, the foetus is very much in command of the pregnancy. It is the foetus who guarantees the endocrine success of pregnancy and induces all manner of changes in maternal physiology to make her a suitable host. . . . It is the foetus who determines the duration of pregnancy. It is the foetus who decides which way it will lie in pregnancy and which way he will present in labour. Even in labour, the foetus is not entirely passive—neither the toothpaste in the tube nor the cork in the champagne bottle, as required by the old hydraulic theories of the mechanics of labour. Much of the behaviour of the neonate and infant can now be observed *in utero* and, by corollary, a better understanding of the foetus and his environment puts the behaviour and problems of the neonate in better perspective. . . . In his warm and humid microclimate, the foetus is in neither stupor nor hypoxic coma. . . . The foetus has been moving his limbs and trunk since about eight weeks, but some 10 or more weeks elapse before these movements are strong enough to be

transmitted to the abdominal wall. . . . Foetal comfort determines foetal position. . . . He will repeatedly and purposefully seek to avoid the sustained pressure of a microphone or phonendoscope or of a knuckle on prominences. . . . The foetus is responsive to pressure and touch. . . . The foetus responds with violent movement to needle puncture and to the intramuscular or intraperitoneal injection of cold or hot hypertonic solutions. . . . Perhaps nowhere does the notion of foetal life as a time of quiescence, or patient and blind development of structures in anticipation of a life and function to begin at birth, die harder than in the concept of the pregnant uterus as a dark and silent world. . . . Smythe (1965) at University College Hospital found that flashing lights applied to the maternal abdominal wall produced fluctuations in foetal heart rate. . . . Sudden noise in a quiet room—the dropped gallipot or maternal voice—startles the foetus lined up under an image intensifier, and from at least 25 weeks the foetus will jump in synchrony with the tympanist's contribution to an orchestral performance. . . . This then is our picture of the foetus. He does not live in a padded, unchanging cocoon in the state of total sensory deprivation, but in a plastic, reactive structure which buffers and filters, perhaps distorts, but does not eliminate the outside world. Nor is the foetus himself inert and stuporose, but active and responsive.[12]

Those who claim that a fetus is not a person point to the fact that it certainly does not behave like one, at least in its earlier stages. They also say that the phenomena of identical twinning and recombination suggest that the fetus is not a person for the first 7 to 12 days, since it seems to lack true, functioning unity. Hence, if a person is not present from the beginning, then why not at a later stage? Finally, they argue that a person is essentially one who relates, or can relate to others; loves and is loved by others. But this is not possible for the fetus, for while its mother (and father) may love it (him or her), it is not able, presumably, to love in return.

Even if it were agreed that the fetus from its very beginning was a person, the question could be asked as to whether or not, under some conditions, it would be possible to suppress one life for the well-being of another. But in doing so, do we make one person the *means* to another? One of the generally accepted values of our culture is that persons are not means but ends in themselves. Consequently, it would not be ethical to suppress one life for the well-being of another, since this would make the first one a means to attain some good for the second.

OTHER ETHICAL ISSUES IN GENETIC COUNSELING

Is the physician obliged to inform his patients of the result of a genetic analysis? Robert Murray would hold that an individual has the right *not* to know, if his ignorance of a genetic defect would be in his best interests, ie his well-being, freedom, and happiness.[13] In the light of the foregoing discussion of the doctor-patient relationship, the physician's task is to determine whether imparting or withholding the genetic information would be most beneficial to his patient from the patient's point of view. Some would hold that information is truth and that a person has a right to the truth, especially about himself; that it is more in keeping with the dignity of a person for him to know the truth about himself, even an unpleasant truth, than to be kept in ignorance like a child. Perhaps it is the kind of information concerning which nothing can be done. However, one should also remember that it may be information that would modify the patient's behavior with regard to marriage. Before a genetic analysis is made, it might be advisable for the physician to reach an agreement with the patient as to whether he wants information conveyed to him or wishes to place some restrictions on that information.

Similarly, if it is judged that the patient's self-image may be damaged by telling him that he is a carrier of an affected gene, should he be told? The answer to that might depend in part on the manner in which the information is obtained. If it is the result of involuntary participation in a mass screening program, then it would seem that such data need not be transmitted to the patient, since he did not freely seek it. This would be especially true if there is nothing the patient could do with the information except, perhaps, worry and agonize about it.

Huntington chorea can be taken as an example of the dilemma which may arise in a counseling situation. As Lissy Jarvik asks, "Once we are able to detect asymptomatic carriers, should we tell them? All of them? Some of them? Is the reassurance we would be able to give to those who are not carriers of the gene, that neither they nor their descendants would be afflicted, worth the certainty of doom that the carriers would simultaneously experience?" (See Chapter 1, p. 7.) Granted that the family as a whole may be involved, still the ethical question focuses generally on what course of action would be most beneficial to the patient, because a physician's primary concern is the well-being of his patient. The analysis would need to take the long-range view: Which option would be more likely to lead to the patient's goal? Which would be most consonant with his values? It seems to me that if a person knows for certain that he is under the sure doom of Huntington chorea, he is in a better position, with the aid of appropriate counseling, to make mature adjustments in his life-style: e.g., not get married or bear children and to avoid long-term commitments. In addition, instead of postponing trips and other fulfilling activities, he would be more likely to undertake them sooner if he felt them to be important. Although the initial shock of being informed of his condition is not desirable, nonetheless there seems to be a positive gain if the remaining years of the patient's life can thus be lived constructively.

To what extent does the question of whether to impart privileged communication vary in different situations? Returning to the preceding example, should a physician inform a woman that her prospective husband, his patient, has Huntington chorea? Privileged information is not an absolute, and the physician must weigh the best interests of his patient as well as those of other concerned parties. This is especially true when a third person will be injured by not having the information which the physician possesses. In such a conflict, when is the patient's right to confidentiality overridden by the prevention of injury to another? Among the factors to be considered are the extent of the expected injury, e.g., life threatening, health disrupting, psychologically or socially damaging; the probability of its occurrence; an estimation of the ill effects the revelation will have on the patient; and a risk/benefit comparison, ie the negative effects on the patient versus the positive gains for the third party.

Should the physician then decide to act, it would seem necessary for him first to attempt to persuade his patient to convey that information to the individual who is the subject of potential injury. Failing that, the physician may, depending upon the seriousness of the problem, take it upon himself to relay that information to the other party. However, that step should not be taken lightly. Once confidentiality has been breached, it is almost impossible to restore it in that patient. Furthermore, there is a societal dimension: others who hear about the "information leakage" might wonder about the confidentiality they have with their own physician. This is not an insignificant consideration. Individuals sometimes suffer to protect the well-being of the community, e.g., policemen, firemen, medical personnel working in disaster areas, soldiers. Presumably, they are exposing themselves knowingly and freely to some danger. There is a kind of informed consent. But not so in the illustration being considered. Yet could consent be reasonably presumed if the patient knew all the circumstances? Would he not willingly sustain the injury in order that the institution of confidentiality be safeguarded? There is no easy answer. While the physician may consider these factors and others, the painful decision is still his to make.

A related issue: Could the state require that the results of genetic analysis be registered in a central data bank? Some people would see many threats to confidentiality, along with the larger threat of government control. Computer people are concerned because what man puts into a data bank can also be gotten out, even though precautions may be taken to protect the information. This is a complex question which can only be noted in this chapter. It merits full discussion in another presentation.

Should the physician attempt to *persuade* his patient to take a particular course of action—for example, to abort a defective fetus? In so doing, the physician would seem to be imposing his own moral stance upon his client. In our present atmosphere, where abortion seems to be so readily accepted, the physician could consider the possibility that he has foreclosed other options too quickly. Furthermore, if abortion is seen as a ready, almost universal "solution" to a genetic problem, research into alternative ways of preventing and/or treating such diseases is less likely to be pressed with adequate vigor and funding. The persistent presence of a medical problem has been a great incentive to discoveries in the realm of treatment and prevention.

Notwithstanding the progress that has been made in the prevention and treatment of genetic diseases, some people may hold to the mistaken notion that any inherited genetic disorder is invariably and severely disabling. Metabolic defects, for example, are expressed in a great variety of ways, ranging from an asymptomatic form without consequence all the way to lethality. Between these extremes, the disease can be asymptomatic except for accidental circumstances which trigger its expression; it may be mild to moderate, involving troublesome conditions yet compatible with long life and productivity; or it can be more severe and seriously incapacitating.[14]

Some of these metabolic diseases lend themselves to the employment of certain therapeutic measures, but many others do not. Some of the ways in which the defect can be corrected may involve the following: supplying the missing metabolite (for example, metabolic cretinism can be ameliorated by supplying thyroxine at a sufficiently early age); supplying a vitamin cofactor (homocystinuria responds, as far as the excessive quantity of homocystine is concerned, to pyridoxine); limiting the intake of a precursor which may undergo toxic accumulation (phenylketonuria has been treated with some degree of success by means of diets low in phenylalanine); providing a necessary tissue by organ transplantation (this is possible in certain cases, although the immunologic barriers still present a problem; for example, the transplant of allogenic marrow in patients wih immunologic deficiency states has had some degree of success).

Mindful of this developing field, the physician must proceed cautiously in urging a specific course of action. This seems to be the attitude implied in the definition of genetic counseling given by William S. Sly: "the delivery of professional advice concerning the magnitude of, the implications of, and the alternatives for dealing with the risk of occurrence of a hereditary disorder within a family." [15] Advice is given regarding alternatives. It is more in keeping with the patient's responsible autonomy that he not be "coerced" to follow a particular course personally favored by his physician, who should serve in an advisory capacity, to aid the patient in making a decision. If any direction is to be favored, it should be the one most closely consonant with the patient's value system. In this manner, the patient will have moved more closely toward his own self-fulfillment.

When genetic analysis reveals that a fetus has a genetic disease and the parents elect not to have an abortion, the psychiatrist has the important role of preparing them for the birth of that child. Not only will the parents be emotionally prepared, but they will also be in a better position to take what practical steps are necessary for the "care and feeding" of their child. This may be a difficult task for the psychiatrist who is firmly convinced that abortion was the only reasonable step to take. He may reject the patient outright or, more likely, continue giving his attention to the medical problem but resent the person who has it. Again, I see the physician as his client's agent; consequently, if he is to be of continuing service, he must use his skills to help the parents deal constructively with their problem and prepare them for the likely outcome, the birth of a defective child whose limitations cannot ordinarily be adequately assessed until after birth.

If the parents themselves differ on a course of action after genetic consultation, e.g., to abort or not to abort, who then decides? The physician? Who breaks the tie vote? Obviously, the physician should help the couple to arrive at a mutual decision and work through their differences. But if they fail to do so, then it would seem that the woman's wish should be followed, since she is the one who has the closest contact with the issue.

What about the sense of shame and guilt associated with genetic disease or the failure to produce a healthy child? * It seems to be part of the "American dream" that everyone is expected to produce healthy, beautiful, well-adjusted, and successful offspring. There seems to be no room for variation. One might recall an important aphorism, "Freedom is the right to fail." [16] In this connection, it might be well to note Callahan's observation that Tay-Sachs disease is an abnormality, but the child with that disease is not thereby an unworthy human being.[17] Furthermore, it seems to be a limitation of freedom by classifying as irresponsible the action of the person who carries a defective fetus to term. There is a presumption here which needs to be carefully analyzed. That presumption is that society's standards are necessarily superior to the individual's standards. This leads to the difficult question of the individual versus the community.

As Crow has asked, how responsible is one generation to the next? [18] Or, as Hotchkiss phrased the question, is the gene pool of mankind public property? [19] These formulations pertain to the same issue: the one versus the many. In our culture (at least up to the present), the indi-

* See also Chapter 7, pp. 120–127.

vidual is not to be sacrificed for others without his free, informed consent. On the other hand, there is the problem of the common good. There are times, it seems, when the individual has to submerge his own well-being for the sake of others. This is generally acceptable if the individual does so freely. But when he is coerced, that is another matter. Is this ethically right? With regard to warfare we have recognized, to some extent at least, the rights of the conscientious objector. The presumption is, therefore, that if one is not a conscientious objector, then to some degree one is acquiescing to his participation in the armed services. But in the area of genetics, can the state by legislative action prohibit marriages or reproduction between certain individuals because of their genetic composition, or can the state require that a defective fetus be aborted?

It has taken a long time for us to learn compassion toward the needy, toward those who are less productive, and toward those who, by accident or by birth, are incapable of taking care of themselves. To embark upon a *forced* abortion policy in the case of defective genes would be a backward step, in that we would eliminate the problem by eliminating the person. Furthermore, not every genetic disease is so disabling that the person who has it cannot contribute to society. History is filled with individuals who have made their mark despite a genetic defect. A classic example is Lincoln, who is thought to have had Marfan syndrome.

These are indeed thorny questions, and they are beyond the primary objectives of this discussion. But they are raised here because they are important and because the idea of compulsory abortion is being proposed with increasing frequency. Many societies prohibit marriage if one or both individuals have a venereal disease. But, since in most cases such an illness is readily treated by chemotherapy, this does not represent a significant infringement of the person's right to marriage. In the case of genetic disease, however, the individual would be required to lead a celibate life or, if he married, to have his reproductive functions surgically or chemically impeded. For many persons this procedure does represent a marked reduction of freedom. There is no easy resolution of this difficulty; free and open public debate will be necessary.

The prime involvement of the psychiatrist in the area of genetic counseling would be to help his patient deal with the emotional consequences of learning of the presence of a genetic disease. More is at stake than is at first apparent. It seems a very attractive hypothesis to present genetic improvement of the human race as a necessary prerequisite to improvement of the quality of life. But quality of life should not be judged solely in terms of usefulness or productivity; it should include other characteristics, such as justice, love, freedom, the ability to sustain pain and disappointment, and the ability to live in an imperfect world, one which can, however, be made ever more humanizing. If, together with Leon Kass, we can say that all humans are radically equal in their genetic constitution, then we must treat them accordingly.[20] The psychiatrist in his office may not have to deal, at present, with the far-reaching consequences of genetic engineering or genetic screening, but he will have to deal from time to time with individuals who are confronted with a genetic problem. To help to meet this problem ethically and constructively is one of his tasks.

REFERENCES

1. Hecht, F. and Holmes, L. B.: What we don't know about genetic counseling. *N. Engl. J. Med. 287:*464–465, 1972.

2. Ethical and social issues in screening genetic disease. *N. Engl. J. Med. 286:*1129–1132, 1972.

3. Osmundsen, J. A.: We are all mutants—preventive genetic medicine: a growing clinical field troubled by confusing ethicists. *Medical Dimensions 2:*5–7, passim, 1973.

4. Morison, R. S.: Chairman's introduction. In Harris, M. (ed.): *Early Diagnosis of Human Genetic Defects: Scientific and Ethical Considerations* (Fogarty International Proceedings no. 6). Washington, D.C.: Government Printing Office, 1972, pp. 7–10.

5. Hamilton, M. P. (ed.): *The New Genetics and the Future of Man.* Grand Rapids: Wm. B. Eerdmans, 1972.

6. Diddle, A. W.: Rights affecting human reproduction. *Obstet. Gynecol. 41:*789–794, 1973.

7. Sonneborn, T. M.: Ethical issues arising from the possible uses of genetic knowledge. In Hilton, B., Callahan, D., Harris, M., Condliffe, P., and Berkley, B. (eds.): *Ethical Issues in Human Genetics: Genetic Counseling and the Use of Genetic Knowledge* (Fogarty International Proceedings no. 13). New York: Plenum Press, 1973, pp. 1–6.

8. Danks, D. M.: Prospects for the prevention of genetic disease. *Med. J. Aust. 1:*573–577, 1973.

9. Attwater, D. (ed.): *A Catholic Dictionary,* 3rd ed. New York: Macmillan, 1958.

10. Reich, C. A.: *The Greening of America: The Coming of a New Consciousness and the Rebirth of a Future.* New York: Random House, 1970.

11. Boethius: Liber de persona et duabus naturis contra Eutychem et *Nestorius,* chap. 2. Thomas Aquinas: Summa Theologica III, Q. 16, a. 12 ad 2um.

12. Liley, A. W.: The foetus as a personality. *Aust. N.Z. J. Psychiatry 6:*99–105, 1972.

13. Murray, R.: Screening: A practitioner's view. In Hilton et al (eds.), *Ethical Issues,* pp. 121–130.

14. Stanberry, J. B. (ed.): *The Metabolic Basis of Inherited Disease,* 3rd ed. New York: McGraw-Hill, 1972.

15. Sly, W. S.: What is genetic counseling? In Bergsma, D. (ed.): *Contemporary Genetic Counseling.* Birth Defects: Orig. Art. Ser., vol. 9, no. 4. White Plains: The National Foundation–March of Dimes, 1974, pp. 5–18.

16. Burch, N. R.: Personal communication.

17. Callahan, D.: The meaning and significance of genetic disease: philosophical perspectives. In Hilton et al. (eds.), *Ethical Issues,* pp. 83–90.

18. Crow, J. F.: Population perspective. Ibid., pp. 73–81.

19. Hotchkiss, R. D.: Discussion after F. C. Fraser: Survey of counseling practices. Ibid., pp. 19–20.

20. Kass, L. R.: Implications of prenatal diagnosis for the human right to life. Ibid., pp. 185–199.

11

LEGAL ISSUES IN MEDICAL GENETICS

Margery W. Shaw, M.D.

The science of modern genetics, as it applies to man, is changing the pattern of traditional medical practice and challenging the conservative system of jurisprudence. Physicians have always held their patients' interests to be paramount to all other considerations; lawyers have cherished their clients' rights without question. But geneticists find themselves concerned with something beyond the individual: the generations of mankind yet unborn.

Recent advances in medical genetics have raised legal issues which are without precedent. The courts will be confronted with new forms of legal action. One can only speculate how these challenges will be met. It is the purpose of this chapter to describe several examples of genetic problems which intersect with the law.

Chromosomes and Crime: The XYY Syndrome

The knowledge gathered about the XYY syndrome since 1965 has been reviewed. It can be summarized in the following statements:

(1) The frequency of XYY males at birth is approximately 1:500.[1]
(2) The frequency of XYY males in selected institutionalized groups such as prisons, hospitals for the criminally insane, and "violent" psychiatric patients is approximately 1:50.[1]
(3) The XYY male is, on the average, 5 inches taller than his XY counterpart.[2]
(4) XYY males with criminal records have a mean age of 13.1 years at first offense, compared to 18.0 for XY males.[3]
(5) Only one of 31 brothers of XYY prisoners had a criminal record, compared to 139 convictions among 63 male sibs of XY offenders.[3]
(6) A mean IQ of 89 has been reported among 18 XYY males tested, although some individuals with XYY have an IQ above 120.[3a]

These statistics do not prove a cause-and-effect relationship, but they do point to a strong correlation between the chromosomal abnormality and excess height, mild mental retardation, aggressive tendencies, and criminal behavior. Heller [4] quotes Lejeune as saying, "There are no born criminals but persons with the XYY defect have considerably higher chances."

It was inevitable that the XYY defense would sooner or later be raised in criminal proceedings. In 1968 a Paris court heard the first evidence of the XYY condition in a man being tried for the murder of a prostitute.[5] He was found guilty, but his sentence was reduced because of "diminished responsibility." That same year, an Australian with the XYY syndrome was found not guilty by reason of insanity after he murdered his 77-year-old landlady.[6] The defense argued that "he had an extra Y chromosome in every cell of his brain." [7] He was committed to a mental hospital at the governor's pleasure.[6] In England an XYY male pleaded guilty to manslaughter of his four children. Without raising the issue of his chromosomal status, three psychiatrists testified that he had a long history of psychopathic disorder, including admission to mental hospitals. There was no evidence of psychosis. He was ordered to Broadmoor Hospital for unlimited duration under the Mental Health Act of 1959. The whole proceeding took less than half an hour.[8] Prior to this case, Gibbons, a British forensic psychiatrist, had offered the prediction that "if a man has no detectable abnormality except the XYY . . . it is very unlikely that this would be regarded in the British courts as evidence of diminished responsibility." [9]

At least three cases involving the XYY syndrome have been tried in the United States. In *People* v. *Farley* [10] both a psychiatrist and a geneticist testified for the defense.[11] The jury found the defendant guilty of rape and murder following the prosecution's argument that he was in a "drunken rage" when he committed the murder.[12] The court sentenced him to prison for 25 years to life.[13]

In Los Angeles, a trial judge disallowed the chromosome defense as admissible evidence.[14] He stated that there was no scientific proof that a double dose of Y chromosomes had made the defendant unable to control himself and that allowing this defense might open a "Pandora's box." [15]

In *Millard* v. *State* [16] an XYY defendant was found guilty of robbing with a deadly weapon; he appealed. The Court of Special Appeals of Maryland affirmed the judgment of the lower court. The appellate judge concluded, ". . . we do not intend to hold, as a matter of law, that a defense of insanity based upon the so-called XYY genetic defect is beyond the pale of proof. . . . We only conclude that on the record before us the trial judge properly declined to permit the [XYY evidence] to go to the jury. . . ." Thus, it seems that with this decision the door has been wedged open in American courts for evidence of the XYY defense, and more cases will surely come.

It should also be mentioned that Richard Speck, the infamous Chicago murderer of eight nurses, purportedly carried an extra Y chromosome,[17] although Speck's attorney later announced that his client's chromosomes were normal.[1]

Uthoff [18] describes the four legal tests which may be applied by the court in an XYY case: (1) M'Naghten's Rule,[19] (2) the Irresistible Impulse Test,[20] (3) the Durham Rule,[21] and (4) the Model Penal Code.[22]

Only the third and fourth are presently useful with our limited state of knowledge of the cause-and-effect relationship of chromosomes and crime. The Durham Rule (or the "Product Test") has included sociopathic personalities in some jurisdictions (i.e., where "his act was the *product* of some mental disease or defect"). The proposed official draft of the Model Penal Code is somewhat more inclusive, stating "if he lacks the capacity to conform his conduct due to a mental disease or defect." It is obvious that the locale or jurisdiction will play an important role in the outcome of these cases.

Housley [23] points out that the courts have accepted radar tests, drunkometer tests, and blood tests, and that the law has recognized other physical conditions which cause aberrant conduct, such as epilepsy, metabolic dysfunctions, and encephalitis. If the XYY condition were classified under "organic brain syndrome," it might be easier for the courts to cope with it.

Turman [13] compares the XYY syndrome to other physical diseases. He suggests that perhaps there is a parallel to be drawn to the narcotics addict or the chronic alcoholic. In *Robinson* v. *California,*[24] the Supreme Court ruled that to punish an addict for his disease is unconstitutional because it violates the Eighth Amendment safeguard against cruel and unusual punishment.

Saxe [25] pleads for better understanding between science and the law. Any sound legal system must possess the capacity to grow and adapt itself to the surrounding environment of increased scientific knowledge.

An editorial by Elkington [26] raises several cogent questions. "Does the presence of an extra Y chromosome lessen the ethical and legal responsibility of an individual for his acts? If it does not mitigate his guilt, should it modify the nature of his sentence? Does it limit his freedom to make responsible choices? At what point does an antisocial person with abnormal chromosomes cease to be a criminal under the jurisdiction of the law and become a patient under the care of the medical profession?" Many of these questions have not yet been addressed to the courts, and there are no simple answers to them.

Bazelon [27] states that the legal question is whether the XYY man is responsible for his crime. A French court said yes; an Australian court said no. The simplest solution would be to make the evidence admissible and present it to the jury. But the jurors will only be confused unless we can clarify for them the relationship between XYY and blameworthiness. Our present state of scientific knowledge does not permit us to do so.

Thus, it is perhaps better that it be a judge's decision—but he also is in a dilemma. If he allows insanity as a defense, then it is arguable that the insanity is permanent and incurable, since "every cell of the body is affected" and there is no known treatment of chromosomal disease. This is a weak argument, however. Diabetes and blindness are also incurable, but we can treat the diabetic and we can help the blind to adjust to their handicap. The medical and legal professions should establish a firm policy of treating the criminal rather than the crime.

Mating Prohibition and Sterilization on Eugenic Grounds

The term "eugenics" means the improvement of the human race by selective breeding. Positive eugenics refers to the promotion of racial betterment by encouraging reproduction by those considered "fit," while negative eugenics applies to prevention of the reproduction of those regarded as "unfit." [28]

During the first three decades of this century, a great American social movement aimed at applying the principles of animal and plant breeding to man. There was one major difference, however; crop and stock improvement was done principally by positive eugenic principles, whereas in man negative eugenics was more often applied. This culminated in a series of laws which were passed in many states in an attempt to eliminate "harmful" genes before the carrier reached reproductive age. These statutes involved prohibitions against certain types of marriage, such as incestuous and interracial marriage, and for sterilization of epileptics, mental defectives, sex offenders, and habitual criminals.[29] Although the motives of the legislators may have been laudable, their efforts were based on misinformation about genetics. Fortunately, most of these laws have been erased from the books, by amendment, repeal, or court decision. But it was not until 1967, in *Loving* v. *Virginia,*[30] that the United States Supreme Court struck down a Virginia statute which prohibited interracial marriage. The laws on incest and consanguinity have seldom been challenged in the courts, although prohibitions against affinous relationships (e.g., stepparent-stepchild) have no genetic basis.

The Supreme Court has also handed down two leading decisions on sterilization. In *Buck* v. *Bell,*[31] the Court upheld a Virginia statute which authorized the sterilization of mental defectives when the defect was shown to be hereditary. Justice Oliver Wendell Holmes made the famous remark that "three generations of imbeciles is enough." In *Skinner* v. *Oklahoma*,[32] the Supreme Court took a different approach and held that an Oklahoma statute providing for the compulsory sterilization of habitual criminals violated the Equal Protection Clause of the Fourteenth Amendment. In both these cases the justices lacked scientific information concerning the hereditary nature of the condition in question. In the *Bell* case the Court assumed it was hereditary; in the *Skinner* case they did not reach the issue.

Genetic Screening

Genetic screening is the process of surveying target populations for genetic disease or for the presence of abnormal genes in the heterozygous condition (the "carrier" state). The populations involved may be screened at different ages, or they may be members of high-risk groups, such as families in which the disease in question has already occurred, religious isolates, or particular ethnic groups.

Over 80 tests are now available for prenatal screening by amniocentesis. Screening of newborn populations has been done for inborn errors of metabolism, hemoglobinopathies, and chromosomal abnormalities. Schoolchildren are particularly easy to reach as a cohort sample. Premarital and postmarital screening is also popular.

Many of the screening programs have been conducted as pilot studies by geneticists who wished to evaluate certain gene frequencies in selected groups. These have been on a voluntary basis. But in some states screening has become mandatory by statute for certain diseases. The most widespread examples are the phenylketonuria (PKU) neonatal screening laws enacted by all 50 states,[33] although in seven of them the laws were repealed after the case-finding costs were found to be prohibitive in relation to societal benefits.

PKU is an autosomal recessive disease which produces profound mental retardation. In recessive inheritance, both parents are "carriers" of the abnormal gene, and there is one chance in four that any offspring will be affected. PKU screening laws are based on sound genetic evidence. Furthermore, not only are new cases identified but a special diet is also available which prevents, partially or entirely, the development of mental retardation if treatment is begun during the first few weeks of life.

Sickle cell screening legislation has become a very popular political tool because of its potential advantages to blacks. By March 1973, 12 states and the District of Columbia had passed laws on screening for this disease.[34] Reilly [35] has analyzed the defects in these laws, which were caused, partially at least, by lack of understanding of the disease itself and of the nature of its genetic transmission. The heterozygous "trait" carrier and the homozygous "affected" individual were confused in some cases. Most of the states made screening mandatory in specific situations, and only one state provided for confidentiality of the test results. Provisions for counseling and education were lacking in most statutes.

In contrast to sickle cell anemia, a very carefully planned voluntary screening program for Tay-Sachs carrier detection was carried on in the synagogues of the Baltimore-Washington area for young married couples (Ashkenazic Jews have a much higher frequency for the Tay-Sachs gene than do non-Jews). Here the results were followed by counseling, and if pregnancy ensued, amniocentesis was offered as a diagnostic test on the fetus.[36] This project emphasized the opportunity for all Jewish couples at risk to "plan" their families without fear of affected children.

As more biochemical tests are discovered, other diseases and carrier states will undoubtedly be screened. Of particular interest to psychiatrists would be the development of a prenatal test for Wilson disease or Huntington chorea. These diseases follow simple Mendelian inheritance. Wilson disease is due to an autosomal recessive condition, while Huntington chorea is an autosomal dominant with late onset. Most recently, some forms of manic-depressive psychosis have been related to X-linked inheritance.[37] Prenatal screening for the sex of the fetus would offer the hope of avoiding the disease in the offspring, but by this method normal fetuses would be sacrificed also.

Genetic Counseling

To return to the PKU tests, it would appear that there are no moral, ethical, or legal issues involved in screening and treatment of this disease. However, the geneticist is concerned lest the frequency of the PKU gene increase in future generations because, as a direct result of medical intervention, more of the affected individuals grow up and reproduce. The physician shoulders this responsibility: is it his ethical duty to counsel these patients when they reach reproductive age? Is it also his legal duty to warn them of the potential risks to future generations? If so, the burden shifts to the individual who is counseled. What is his duty? What are his rights regarding procreation? What are the rights, if any, of the unborn child whose parent *knowingly* transmits to him an abnormal gene? Will the courts hold that invidious discrimination is occurring against a certain group of individuals in violation of the Equal Protection Clause? These questions cannot be answered, but Dr. Bentley Glass put them in perspective when he wrote: "The paramount right of the individual is not the right to reproduce, often denied in past societies; it is the right of the child to be born physically and mentally sound." [38]

Wrongful Life

In the future, if a couple willfully and knowingly conceives children when both parents have been informed that the risk of genetic disease or defect is high and that the burden of the disease is great, it is not far-fetched to presume that the affected children may sue their parents for "wrongful life." There are precedents for this cause of action.[39] In *Zepeda* v. *Zepeda* [40] an illegitimate child sought damages against his father. Although the Illinois court concluded that the father had, in fact, committed a tort at the time of conception and that the child was harmed by the stigma of his illegitimacy, it refused to award damages for wrongful life. Concerned that there would be a flood of lawsuits on bastardy, the court exercised judicial restraint stating that the remedy should come from state legislation rather than from ad hoc case litigation.

In *Gleitman* v. *Cosgrove,*[41] a congenitally defective child whose mother had rubella during early pregnancy sued for damages in a New Jersey district court for the tort of wrongful life. The child was born with a congenital heart defect and was also deaf and blind. The parents also sued the physician for mental and emotional pain and suffering and economic loss because the obstetrician negligently failed to advise the mother that her child might be defective. In fact, he had offered her reassurances and told her not to worry. The court decided in favor of the physician. On appeal, the Supreme Court of New Jersey held that the child could not recover because if the physician had not been negligent, he would not have been born at all! However, two justices dissented on the question of the duty of the physician to disclose the risks, saying ". . . they were under a clear duty to tell her of its high incidence of abnormal birth. That duty was not only a moral one but a legal one as well."[42] The court also decided that the parents were not entitled to damages, since it was against public policy to take an embryonic life. In other words, the court held that the child's right to life, miserable though it might be, outweighed the parents' rights not to endure emotional suffering or financial injury. In light of the *Roe* v. *Wade* decision in 1973,[43] one wonders if the *Gleitman* outcome would be different now. In *Roe,* the United States Supreme Court held that the mother's decision to have an abortion during the first trimester of pregnancy outweighs the fetus's right to life.

Contraception and Abortion

The landmark decision of *Griswold* v. *Connecticut* in 1965 established the right of privacy with regard to sex and procreation.[44] *Griswold* held Connecticut's birth control statute unconstitutional under the First, Third, Fourth, Fifth, Ninth, and Fourteenth Amendments. The court leaned heavily on the Ninth Amendment, which provides that "the enumeration in the Constitution, of certain rights, shall not be construed to deny or disparage others retained by the people.''

The *Griswold* ruling laid the framework for the Supreme Court abortion decision in *Roe* v. *Wade* in 1973.[43] In *Roe,* the Court held that in the first trimester of pregnancy the decision to abort is primarily the concern of the patient and her physician; the reasoning relied primarily on the *Griswold* precedent of right to privacy. In the second trimester, the state has an interest in controlling the place and method of abortion, but no overriding interest in protecting the fetus's rights over those of the mother. It is only in the last trimester, when the fetus becomes viable outside the womb, that the state may proscribe abortion except to preserve the life and health of the mother.

Critics have attacked the holding in *Roe* because of its arbitrariness. Scientists would agree that advances in technology will surely change the definition of extrauterine viability, casting doubt on the basis for the Court's reasoning. Another criticism is that interference with conception is not in the same category as interference with fetal survival, and that the newly created being has rights which should be weighed against those of the mother. Some argue that these rights should begin at the moment of conception. An inescapable conclusion of the latter argument is that if the life of the zygote is sacred, then there is no greater wrongdoing in infanticide than in interruption of early pregnancy. Yet most people would feel such a practice to be a greater affront to human dignity than early abortion. The characteristics of ''personhood'' develop slowly and imperceptibly after conception, but a human liveborn infant seems to most of us to be much more ''human'' than a fertilized egg or a blastocyst.

Another basic legal concept which can be applied to abortion is the ''right to control one's own body.'' Justice Brandeis articulated the forerunner of this concept in 1928 in his famous remark, ''a right to be let alone—the most comprehensive of rights and the right most valued by civilized man.'' [45] The recent decisions on informed consent [46–48] stress that the physician's duty to disclose knowledge about the risks and alternatives of treatment is premised on the basic legal dictum that ''every human being of adult years and sound mind has a right to determine what shall be done with his own body. . . .'' [49] But no rights are absolute; personal rights are protected only if their exercise does not cause harm to others. In the case of genetic disease, one's right to reproduce may be questioned if it inflicts pain and suffering on one's offspring.

REFERENCES

1. Shah, S. A.: *Report on the XYY Chromosomal Abnormality*. Center for Studies of Crime and Delinquency. Rockville, MD, National Institute of Mental Health. Public Health Service Publication, 1970, pp. 1–55.

2. Court Brown, W. M.: Males with an XYY sex chromosome complement. *J. Med. Genet. 5:*341, 1968.

3. Price, W. H. and Whatmore, P. B.: Behavior disorders and the pattern of crime among XYY males identified at a maximum security hospital. *Br. Med. J. 1:*533–536, 1967.

3a. Money, J., Gaskin, R. J., and Hull, H.: Impulse, aggression and sexuality in the XYY syndrome. *St. John's Law Rev. 44:*220–235, 1969.

4. Heller, J. H.: Human chromosome abnormalities as related to physical and mental dysfunction. *J. Hered. 60:*239–248, 1969.

5. *Medical Tribune,* November 4, 1968, sec. 1, p. 26.

6. *Time,* October 25, 1968, p. 76.

7. Abnormal chromosome. *Br. Med. J. 4:*398, 1968.

8. An English XYY murder trial. *Br. Med. J. 1:*201, 1969.

9. Gibbons, T.: Chromosomes and abnormal behavior. *Medical-Moral Newsletter 5:*10–11, 1968.

10. *People* v. *Farley,* appeal filed no. 2061, App. Div., 2d, July 24, 1969.

11. Farrell, P. T.: The XYY syndrome in criminal law: an introduction. *St. John's Law Review 44:*217–219, 1969.

12. *New York Times,* April 16, 1969, p. 54.

13. Turman, H.: The XYY syndrome: a challenge to our system of criminal responsibility. *New York Law Forum 16:*232–262, 1970.

14. Robitscher, J.: Medical limits of criminality. *Ann. Intern. Med. 73:*849–851, 1970.

15. Getze, G.: Judge rules out chromosome link to behavior. *Philadelphia Inquirer,* March 9, 1969, pp. 22–24.

16. *Millard* v. *State,* 261 A 2d 227–232 (1970).

17. Burke, K. J.: The XYY syndrome: genetics, behavior, and the law. *Denver Law J. 46:*261–284, 1969.

18. Uthoff, G. S.: the XYY chromosome complement: brief applications to criminal insanity tests. *St. Louis L. J. 14:*297–309, 1969.

19. *M'Naghten's Case,* 8 Eng. Rep. 718 (H. L. 1843).

20. Keedy, H.: Irresistible impulse as a defense in the criminal law. *U. Pa. L. Rev. 100:*956–988 (1952).

21. *Durham* v. *U.S.* 237 F 2d 760 (D.C. Cir. 1956).

22. Model Penal Code, Sec. 4.01 (Proposed Official Draft 1962).

23. Housley, R.: Criminal law: the XYY chromosome complement and criminal conduct. *Oklahoma Law Review 22:*287–301, 1969.

24. *Robinson* v. *California* 370 U.S. 660 (1962).

25. Saxe, D. B.: Psychiatry, sociopathy and the XYY chromosome syndrome. *Tulsa Law. J. 6:*243–256, 1970.

26. Elkington, J. R.: Sex chromosomes and crime. *Ann. Intern. Med. 69:*399–401, 1968.

27. Bazelon, D. L.: Medical progress and the legal process. *Pharos 32:*34–40, 1969.

28. Ludmerer, K. M.: *Genetics and American Society: An Historical Appraisal.* Baltimore: Johns Hopkins Press, 1972, chap. 2.

29. Ferster, E. Z.: Eliminating the unfit—is sterilization the answer? *Ohio State L.J. 27:*591, 1966.

30. *Loving* v. *Virginia,* 388 U.S. 1 (1967).

31. *Buck* v. *Bell,* 274 U.S. 200 (1927).

32. *Skinner* v. *Oklahoma,* 316 U.S. 535 (1942).

33. Swazey, J. P.: Phenylketonuria: a case study in biomedical legislation. *J. Urban L. 48:* 883, 1971.

34. Reilly, P.: Sickle cell anemia legislation. *J. Legal Med. 1*(4):39–48, 1973.

35. Reilly, P.: Sickle cell anemia legislation. *J. Legal Med. 1*(5):36–40, 1973.

36. Kaback, M. M.: Discussion. In Bergsma, D. (ed.): *Intrauterine Diagnosis.* Birth Defects: Orig. Art. Ser., vol. 7, no. 5. White Plains: The National Foundation—March of Dimes, 1971, p. 35.

37. Winokur, G.: Genetic findings and methodological considerations in manic depressive disease. *Br. J. Psychol. 117:*267–274, 1970.

38. Glass, B.: The goals of human society. *Bioscience 22:*137, 1972.

39. Tedeschi, G.: Tort liability for wrongful life. *J. Family L. 7:*465, 1967.

40. *Zepeda* v. *Zepeda,* 41 Ill. App. 2d 240 (1963).

41. *Gleitman* v. *Cosgrove,* 49 N.J. 22 (1966).

42. *Gleitman* v. *Cosgrove,* 49 N.J. 49 (Jacobs, J., dissenting).

43. *Roe* v. *Wade,* 410 U.S. 113 (1973).

44. *Griswold* v. *Connecticut,* 381 U.S. 479 (1965).

45. *Olmstead* v. *United States,* 277 U.S. 438, 478 (1928) (Brandeis, J., dissenting).

46. *Cobbs v. Grant,* 502 P 2d 1 (1972).

47. *Canterbury* v. *Spence,* 464 F 2d 772 (1972).

48. *Wilkinson* v. *Vesey,* 295 A 2d 676 (1972).

49. *Schloendorff* v. *Society of New York Hospital,* 105 N.E. 92, 93 (1914).

INDEX